SAFETY AND RELIABILITY IN THE OIL AND GAS INDUSTRY

A PRACTICAL APPROACH

SAFETY AND RELIABILITY IN THE OIL AND GAS INDUSTRY
A PRACTICAL APPROACH

B.S. DHILLON

CRC Press
Taylor & Francis Group
Boca Raton London New York

CRC Press is an imprint of the
Taylor & Francis Group, an **informa** business

CRC Press
Taylor & Francis Group
6000 Broken Sound Parkway NW, Suite 300
Boca Raton, FL 33487-2742

First issued in paperback 2019

© 2016 by Taylor & Francis Group, LLC
CRC Press is an imprint of Taylor & Francis Group, an Informa business

No claim to original U.S. Government works

ISBN-13: 978-1-4987-4656-4 (hbk)
ISBN-13: 978-0-367-87529-9 (hbk)

Visit the Taylor & Francis Web site at
http://www.taylorandfrancis.com

and the CRC Press Web site at
http://www.crcpress.com

This book is affectionately dedicated to the memory of my

late colleague, Professor Daniel J. Gorman.

Contents

Preface

Billions of dollars are spent every year in the area of oil and gas to design, construct/manufacture, operate, and maintain various types of equipment/ systems around the globe. Today, the oil industry is one of the most powerful branches in the world economy. Each year, more than 4 billion metric tons of oil is produced globally.

Needless to say, safety and reliability in the oil and gas industry has become an important issue due to various types of accidents and problems over the years. For example, in 1988, Piper Alpha offshore disaster in the United Kingdom resulted in 167 deaths and total insured loss of about $3.4 billion.

Over the years, a large number of journal and conference proceedings articles on safety and reliability in the oil and gas industry have appeared, but to the best of the author's knowledge, there is no book that covers both these topics within its framework. This poses a substantial obstacle for information seekers on the subject, because they have to consult a number of different and diverse sources.

Thus, the main objective of this book is to combine safety and reliability in the oil and gas industry into a single volume, eliminate the need to consult a number of different and diverse sources in obtaining desired information, and provide up-to-date information on the subject. The sources of most of the material presented are given in the reference section at the end of each chapter. These will be useful to readers if they desire to delve deeper into a specific area. This book contains a chapter on mathematical concepts and another chapter on safety and reliability basics considered useful to understand the contents of subsequent chapters. Furthermore, another chapter is devoted to methods considered useful to perform safety and reliability analyses in the oil and gas industry.

The topics covered in this book are treated in such a manner that the reader will require no previous knowledge to understand the contents. At appropriate places, this book contains examples along with their solutions, and at the end of each chapter, there are numerous problems to test the reader's comprehension in the area. An extensive list of publications dating from 1928 to 2014, relating directly or indirectly to safety and reliability in the oil and gas industry, is provided at the end of this book to give readers a view of the intensity of developments in the area.

This book consists of 11 chapters. Chapter 1 presents the need for and historical developments in safety and reliability in the oil and gas industry; oil and gas industry safety and reliability-related facts, figures, and examples; important terms and definitions; useful sources for obtaining information on safety and reliability in oil and gas industry; and the scope of this book. Chapter 2 reviews mathematical concepts considered useful in understanding subsequent chapters. Some of the topics covered in this chapter are Boolean algebra laws, probability properties, probability distributions, and useful mathematical definitions.

Chapter 3 presents introductory aspects of safety and reliability. Chapter 4 presents a number of methods considered useful for performing safety and reliability analyses in the oil and gas industry. These methods are root cause analysis, hazard and operability analysis, technique of operations review, interface safety analysis, preliminary hazard analysis, job safety analysis, failure modes and effect analysis, fault tree analysis, and the Markov method.

Chapter 5 is devoted to safety in offshore oil and gas industry. Some of the topics covered in this chapter are offshore industrial sector risk picture, offshore worker situation awareness concept, offshore industry accident reporting approach, and offshore industry accidents' case studies. Chapter 6 is devoted to case studies of oil tanker spill-related accidents, oil tanker spill analysis, and oil spill causes. Chapter 7 presents various important aspects of human factors contribution to accidents in the oil and gas industry and fatalities in the industry. Some of the topics covered in this chapter are human factors that affect safety in general, categorization of accident-related human factors in the industrial sector, categories of human factors accident causation in the oil industry, and recommendations to reduce fatal oil and gas industry incidents. Chapter 8 is devoted to case studies of maintenance influence on major accidents in the oil and gas industry and safety-instrumented systems and their spurious activation in the oil and gas industry.

Chapter 9 presents various important aspects of oil and gas industry accident data and accident data analysis. Some of the topics covered in this chapter are oil and gas industry accident databases, onshore and offshore oil and gas industry accident data analysis, and offshore oil and gas rigs accident analysis. Chapter 10 is devoted to oil and gas industry equipment reliability and covers topics such as oil and gas pipeline fault tree analysis, corrosion-related failures, optical connector failures, and fatigue damage initiation assessment in oil and gas steel pipes.

Finally, Chapter 11 presents a total of seven mathematical models for performing various types of safety and reliability analyses in the oil and gas industry.

This book will be useful to many individuals, including professionals working in the area of oil and gas, researchers, instructors, and graduate and senior undergraduate students in the area of engineering, safety and reliability professionals, and engineers at large.

I am deeply indebted to many individuals, including friends, colleagues, and students, for their invisible inputs. The invisible contributions of my children are also appreciated. Last but not least, I thank my wife, Rosy, my other half and friend, for typing this entire book and for her timely help in proofreading.

B.S. Dhillon
Ottawa, Ontario, Canada

Author

B.S. Dhillon is a professor of engineering management in the Department of Mechanical Engineering at the University of Ottawa. He has served as a chairman/director of Mechanical Engineering Department/Engineering Management Program for more than 10 years in the same institution. He is the founder of the probability distribution named *Dhillon distribution/law/model* by statistical researchers in their publications around the world. He has published over 380 (229 [70 single-authored + 159 coauthored] journal and 151 conference proceedings) articles on reliability engineering, maintainability, safety, engineering management, etc. He is and has been on the editorial boards of 12 international scientific journals. In addition, Professor Dhillon has written 43 books on various aspects of health care, engineering management, design, reliability, safety, and quality that have been published by Wiley (1981), Van Nostrand (1982), Butterworth (1983), Marcel Dekker (1984), Pergamon (1986), etc. His books are being used in more than 100 countries and many of them are translated into languages such as German, Russian, Chinese, and Persian (Iranian).

Professor Dhillon has served as the general chairman of two international conferences on reliability and quality control held in Los Angeles and Paris in 1987. He has also served as a consultant to various organizations and bodies and has many years of experience in the industrial sector. He has taught reliability, quality, engineering management, design, and related areas for more than 35 years at the University of Ottawa. He has also lectured in more than 50 countries, including delivering keynote addresses at various international scientific conferences held in North America, Europe, Asia, and Africa. He was a distinguished speaker at the Conference/Workshop on Surgical Errors (sponsored by White House Health and Safety Committee and Pentagon), held at the Capitol Hill (One Constitution Avenue, Washington, DC) in March 2004.

Professor Dhillon attended the University of Wales where he earned a BS in electrical and electronic engineering and an MS in mechanical engineering. He earned a PhD in industrial engineering from the University of Windsor.

1

Introduction

1.1 Background

Oil and gas have played quite an important role throughout world history. Ancient cultures made use of crude oil as a substance to bind materials and as a sealant to waterproof various surfaces [1]. However, the modern oil and gas industry was basically born in the nineteenth century. For example, in the early years of 1800s, merchants constructed dams that allowed oil to float to the waters' surface in the Oil Creek area of Western Pennsylvania [1,2]. Furthermore, in the mid-1850s, the invention of the kerosene lamp led to the establishment of the first U.S. oil company called Pennsylvania Rock Oil Company.

Currently, oil accounts for a large percentage of the world's energy consumption, ranging from as high as 53% for Middle East to as low as 32% for Europe and Asia [3]. Billions of dollars is spent every year in the oil and gas industry to design, construct/manufacture, operate, and maintain various types of systems/equipment around the globe. Safety and reliability in the oil and gas industry have become important issues due to various types of accidents and problems over the years. For example, in 1988, Piper Alpha offshore disaster in the United Kingdom resulted in 167 deaths and a total insured loss of approximately $3.4 billion [4,5].

Since 1928, a large number of publications directly or indirectly related to safety and reliability in the oil and gas industry have appeared. A list of 190 such publications is provided in the appendix.

1.2 Oil and Gas Industry Safety and Reliability-Related Facts, Figures, and Examples

Some of the facts, figures, and examples directly or indirectly concerned with safety and reliability in the oil and gas industry are as follows:

- Currently, about 30% of the world oil and gas production comes from offshore, and in 2005, 2.42 billion metric tons of oil were shipped by tankers [6,7].

- As per the findings of the International Association of Oil and Gas Producers, the number of fatalities increased from 65 in 2011 to 88 in 2012 [8].

- As per the findings of the International Tanker Owners Pollution Federation (ITOPF), during the period 1984–2008, 9351 accidental spills have occurred [9].

- As per References 4 and 10, in 1996, the direct cost of corrosion-related failures, including maintenance in the U.S. petroleum industry, was $3.7 billion per year.

- In the 1988 explosion in shell oil refinery in Norco, Louisiana, hydrocarbon gas escaped from a corroded pipe in a catalytic cracker and was ignited. This resulted in seven deaths and 42 injuries of workers, and its estimated cost was $706 million [11,12].

- Over the past 40 years, up to 2006, 242 accidents of storage tanks have occurred in industrial facilities around the globe, and 74% of such accidents occurred in petroleum refinery, oil terminals, or storage [13].

- During the period 1997–2003, one fatality occurred every 10 days in the U.S. upstream oil and gas industry, and the U.S. exploration and production industry experienced 254 fatalities onshore [14].

- In 1988, an explosion and resulting fire on a North Sea oil production platform named Piper Alpha in the United Kingdom killed 167 persons [4,5,12].

- In 2005, a gas explosion in a shopping center in the northern Russian town of Ukhta killed 17 and injured 17 persons [12].

- Since 1970, by considering oil tanker spills 7 tons and above, approximately 5.74 million tons of oil were lost as a result of oil tanker-related incidents around the globe [15].

- A gas explosion in a Turkish coal mine killed 263 workers near the Black Sea port of Zonguldak in 1992 [12].

- In 2005, an explosion at a British Petroleum (BP) refinery in Texas City, Texas killed 15 and injured over 100 persons [12,16].

- As per Reference 15, since 1970, over 1800 oil tanker spills, greater than 7 tons per spill, have occurred round the globe.

- In 1993, an oil tanker named "Braer" carrying 84,700 tons of Norwegian Gullfaks crude oil and about 1500 tons of heavy bunker oil following its engine failure ran aground in severe weather conditions and lost its entire cargo of oil [17].

- The San Bruno pipeline explosion that occurred in a suburb of San Francisco in 2010 produced a ball of fire 1000 feet high due to the rupture of the natural gas pipeline and killed eight persons [12,18].
- A 72-car train carrying crude oil derailed its 63 cars in 2013 and caused fire and explosions in downtown Lac-Megantic (in Quebec, Canada) that killed 47 persons and destroyed 30 buildings [12].
- In 2002, an oil tanker named "Prestige" carrying 77,000 tons of heavy fuel oil suffered hull damage in heavy seas off northern part of Spain and released about 63,000 tons of its cargo into the sea [19–21].
- A large oil pipeline explosion and the resulting fire in the city of San Martin Texmelucan de Labastida, Puebla, Mexico in 2010 killed 29 and injured 52 persons [12].

1.3 Terms and Definitions

This section presents some useful terms and definitions directly or indirectly concerned with safety and reliability in the oil and gas industry [22–26]:

- *Accident*: An unplanned and undesired event.
- *Continuous task*: A task that involves some kind of tracking activity (e.g., monitoring a changing condition).
- *Downtime*: The time period during which the item/system is not in a condition to perform its specified mission.
- *Fail-safe*: The failure of an item/item part, without endangering humans or damage to equipment or plant facilities.
- *Failure*: The inability of an item to function within the stated guidelines.
- *Failure mode*: The abnormality of an item/system's performance that causes the item/system to be considered as failed.
- *Hazard control*: A means of reducing the risk of exposure to a perceived hazard.
- *Human error*: The failure to perform a specified task (or the performance of a forbidden action) that leads to disruption of scheduled operations or results in damage to equipment and property.
- *Human factors*: A study of the interrelationships between humans, the tools they use, and the surrounding environment in which they work and live.
- *Injury*: A wound or other specific damage.

- *Maintenance*: All the actions necessary for retaining an item/equipment in, or restoring it to, a stated condition.
- *Mission time*: The element of uptime that is required to perform a specified mission profile.
- *Redundancy*: The existence of more than one means to perform a specified function.
- *Reliability*: The probability that an item will carry out its stated mission satisfactorily for the stated time period when used according to the specified conditions.
- *Reliability model*: A model for assessing, predicting, or estimating reliability.
- *Safeguard*: A barrier, device, or procedure developed for the protection of humans.
- *Safety*: Conservation of human life and the prevention of damage to items as per mission-specified requirements.
- *Safety management*: The accomplishment of safety through the efforts of others (i.e., people).
- *Unsafe act*: An act that is not safe for an employee/individual.
- *Unsafe condition:* Any condition, under the right set of conditions that will lead to an accident.

1.4 Useful Sources for Obtaining Information on Safety and Reliability in the Oil and Gas Industry

There are many sources for obtaining information, directly or indirectly, concerned with safety and reliability in the oil and gas industry. Some of the sources considered most useful are presented below under a number of distinct categories.

1.4.1 Journals

- *Accident Analysis and Prevention*
- *Chemical Engineering Transactions*
- *Computers and Industrial Engineering*
- *Energy Policy*
- *Engineering and Technology*
- *Ergonomics*
- *Human Factors and Ergonomics in Engineering*

- *IEEE Industry Applications Magazine*
- *International Journal of Services, Technology, and Management*
- *International Journal of Technology, Policy, and Management*
- *Journal of Disaster Research*
- *Journal of Failure Analysis and Prevention*
- *Journal of Loss Prevention in the Process Industries*
- *Journal of Natural Gas Science and Engineering*
- *Journal of Petroleum Technology*
- *Journal of Quality in Maintenance Engineering*
- *Measurement Techniques*
- *Natural Gas Industry*
- *Oil & Gas Journal*
- *Process Safety and Environmental Protection*
- *Professional Safety*
- *Reliability Engineering and System Safety*
- *Risk Analysis*
- *Risk, Decision, and Policy*
- *Safety Science*
- *Technology in Society*

1.4.2 Conference Proceedings

- *Proceedings of the Abu Dhabi International Petroleum Exhibitions and Conferences*
- *Proceedings of the Annual Conferences on Oil Pollution*
- *Proceedings of the Annual Petroleum and Chemical Industry Conference*
- *Proceedings of the Annual Reliability and Maintainability Symposium*
- *Proceedings of the European Safety and Reliability Conferences: Advances in Safety, Reliability, and Risk Management*
- *Proceedings of the Human Factors and Ergonomics Society Annual Meetings*
- *Proceedings of the International Conference on Health, Safety, and Environment in Oil and Gas Exploration and Production*
- *Proceedings of the International Conferences on Engineering, Science, Construction, and Operations in Challenging Environments*
- *Proceedings of the International Conferences on Ocean, Offshore, and Arctic Engineering*
- *Proceedings of the International Petroleum Technology Conferences*

- *Proceedings of the SPE Asia Pacific Oil and Gas Conferences and Exhibitions*
- *Proceedings of the SPE Latin American and Caribbean Health, Safety, Environment, and Social Responsibility Conferences*
- *Proceedings of the System Safety Conferences*

1.4.3 Books

- Cox, R.F., Walter, M.H., editors, *Offshore Safety and Reliability*, Elsevier Applied Science, London, 1991.
- Cox, S.J., *Reliability, Safety, and Risk Management: An Integrated Approach*, Butterworth-Heinemann, New York, 1991.
- Dhillon, B.S., *Design Reliability: Fundamentals and Applications*, CRC Press, Boca Raton, Florida, 1999.
- Dhillon, B.S., *Engineering Safety: Fundamentals, Techniques, and Applications*, World Scientific Publishing, River Edge, New Jersey, 1996.
- Dhillon, B.S., *Human Reliability: With Human Factors*, Pergamon Press, New York, 1986.
- Dhillon, B.S., *Reliability, Quality, and Safety for Engineers*, CRC Press, Boca Raton, Florida, 2005.
- Goetsch, D.L., *Occupational Safety and Health*, Prentice-Hall, Englewood Cliffs, New Jersey, 1996.
- Graham, J.H., Editor, *Safety, Reliability, and Human Factors*, Van Nostrand Reinhold Company, New York, 1991.
- Handley, W., *Industrial Safety Handbook*, McGraw-Hill Book Company, New York, 1969.
- *Oil and Gas Employee Safety Handbook*, J.J. Keller and Associates, Inc., Neenah, Wisconsin, 2013.
- Salvendy, G., Editor, *Handbook of Human Factors and Ergonomics*, John Wiley and Sons, New York, 2006.
- Winston Revie, R., Editor, *Oil and Gas Pipelines: Integrity and Safety Handbook*, John Wiley and Sons, New York, 2015.

1.4.4 Data Sources

- American National Standards Institute (ANSI), 11 W. 42nd St., New York, 10036.
- Computer Accident/Incident Report System, System Safety Development Center, EG&G, PO Box 1625, Idaho Falls, Ohio.
- Government/Industry Data Exchange Program (GIDEP), GIDEP Operations Center, U.S. Department of the Navy, Corona, California.

- International Occupational Safety and Health Information Center Bureau, International du Travail, CH-1211, Geneva 22, Switzerland.
- National Technical Information Service (NTIS), United States Department of Commerce, 5285 Port Royal Road, Springfield, Virginia.
- Reliability Analysis Center, Rome Air Development Center (RADC), Griffis Air Force Base, Rome, New York.
- Safety Research Information Service (SRIS), National Safety Council, 444 North Michigan Avenue, Chicago, Illinois.

1.4.5 Standards

- DEF-STD-00-55-1, Requirements for Safety-Related Software in Defense Equipment, Department of Defense, Washington, DC.
- IEC 60950, Safety of Information Technology Equipment, International Electro-Technical Commission, Geneva, Switzerland, 1999.
- MIL-STD-721, Definitions of Terms for Reliability and Maintainability, Department of Defense, Washington, DC.
- MIL-STD-756, Reliability Modeling and Prediction, Department of Defense, Washington, DC.
- MIL-STD-785, Reliability Program for Systems and Equipment, Development and Production, Department of Defense, Washington, DC.
- MIL-STD-882, Systems Safety Program for System and Associated Subsystem and Equipment-Requirements, Department of Defense, Washington, DC.
- MIL-STD-1629, Procedures for Performing Failure Mode, Effects and Criticality Analysis, Department of Defense, Washington, DC.
- MIL-STD-1908, Definitions of Human Factors Terms, Department of Defense, Washington, DC.
- MIL-STD-2155, Failure Reporting, Analysis, and Corrective Action (FRACAS), Department of Defense, Washington, DC.

1.4.6 Organizations

- American Association of Drilling Engineers, PO Box 107, Houston, Texas.
- American Gas Association, 400 North Capitol Street NW, Suite 450, Washington, DC.

- American Society of Safety Engineers, 1800 East Oakton St., Des Plaines, Illinois.
- British Safety Council, 62 Chancellors Road, London, UK.
- Canadian Association of Oilwell Drilling Contractors, 717-7th Avenue SW, Calgary, Alberta, Canada.
- International Association of Oil and Gas Producers, 209–215 Blackfriars Road, London, UK.
- International Energy Agency, 9 Rue de la Federation, 75739 Paris Cedex 15, France.
- International Petroleum Industry Environmental Conservation Association (IPIECA), 5th Floor, 209–215 Blackfriars Road, London, UK.
- Interstate Oil and Gas Compact Commission, 900 NE 23rd Street, Oklahoma City, Oklahoma.
- Occupational Safety and Health Administration, U.S. Department of Labor, 200 Constitution Avenue, Washington, DC.
- Organization of the Petroleum Exporting Countries, Helferstorferstrasse 17, A-1010, Vienna, Austria.
- Reliability Society, IEEE, PO Box 1331, Piscataway, New Jersey.
- U.S. Oil and Gas Association, 1101 K Street NW, Suite 425, Washington, DC.

1.5 Scope of This Book

Each year a vast sum of money is spent around the globe in the oil and gas industrial sector to design, construct/manufacture, operate, and maintain various types of equipment/systems. Safety and reliability in this sector have become an important issue due to various types of accidents and problems over the years.

Over the years, a large number of journal and conference proceedings articles, technical reports, etc., directly or indirectly, on safety and reliability in the oil and gas industry have appeared in the literature. However, to the best of the author's knowledge, there is no book on the topic that covers recent developments in the area. Therefore, this book not only attempts to cover the safety and reliability in the oil and gas industry within its framework, but also provides the latest developments in the area.

Thus, the main objective of this book is to provide professionals and others concerned with safety and reliability in oil and gas industry up-to-date information that could be useful for improving safety and reliability in this area. This book will be useful to many individuals, including professionals working in the area of oil and gas, researchers, instructors, and graduate and senior undergraduate students in the area of engineering, safety and reliability professionals, and engineers at large.

PROBLEMS

1. Write an essay on safety and reliability in the oil and gas industry.
2. Define the following terms:
   - Safety
   - Reliability
   - Safety management
3. List five important facts and figures concerning the safety and reliability in the oil and gas industry.
4. List six important journals for obtaining information related to safety and reliability in the oil and gas industry.
5. What is the difference between the terms unsafe condition and unsafe act?
6. Define the following terms:
   - Human factors
   - Mission time
   - Reliability model
7. List at least four books considered important for obtaining information related to safety and reliability in the oil and gas industry.
8. Compare the terms failure and human error.
9. Define the following terms:
   - Accident
   - Fail-safe
   - Failure mode
10. List at least five standards that can be directly or indirectly concerned with the safety and reliability in the oil and gas industry.

References

1. History of Oil and Gas Industry, retrieved on August 18, 2015 from website: http://www.loc.gov/rr/business/BERA/issue5/history.html.

2. The History of the Oil Industry, San Joaquin Geological Society, retrieved on August 18, 2015 from website: http://www.sjvgeology.org/history/index. html.

3. Petroleum Industry, retrieved on August 18, 2015 from website: https:// en.wikipedia.org/wiki/Petroleum_industry (last modified on January 25, 2016).

4. Pate-Cornell, M.E., Risk analysis and risk management for offshore platforms: Lessons from the Piper Alpha accident, *Journal of Offshore Mechanics and Arctic Engineering*, 115(1), 1993, 179–190.

5. Pate-Cornell, M.E., Learning from the Piper Alpha accident: Analysis of technical and organizational factors, *Risk Analysis*, 13(2), 1993, 215–232.

6. About Offshore Oil and Gas Industry, retrieved on June 19, 2015 from website: http://www:modec.com/about/industry/oil-gas.html.

7. Oil Tanker, retrieved on June 9, 2015 from website: https://en.wikipedia.org/ wiki/Oil_tanker (last modified on January 15, 2016).

8. Safety Performance Indicators: 2012 Data, Report No. 2012, International Association of Oil and Gas Producers, London, June 2013.

9. Oil Tanker Spill Information Pack, retrieved on October 8, 2008 from website: http://www.itopf.com/information-services/data-and-statistics/. International Tanker Owners Pollution Federation, London, UK.

10. Kane, R.D., Corrosion in petroleum refining and petrochemical operations, in *Metals Handbook, Vol. 13C: Environments and Industries,* edited by S.O. Cramer and B.S. Covino, ASM International, Metals Park, Ohio, 2003, pp. 967–1014.

11. Oil Refinery, retrieved on January 13, 2015 from website: https://en.wikipedia. org/wiki/Oil_refinery (last modified on January 19, 2016).

12. Energy Accidents, retrieved on January 13, 2015 from website: https:// en.wikipedia.org/wiki/Energy_accidents (last modified on November 27, 2015).

13. Chang, J.I., Lin, C.C., A study of storage tank accidents, *Journal of Loss Prevention in the Process Industries*, 19, 2006, 51–59.

14. Denney, D., Strategic direction for reducing fatal oil and gas industry incidents, *Journal of Petroleum Technology*, 2005, 66–68.

15. Oil Tanker Spill Statistics 2013, Report, The International Tanker Owners Pollution Federation (ITOPF) Limited, London, UK, 2014.

16. List of Industrial Disasters, retrieved on January 13, 2015 from website: https://en.wikipedia.org/wiki/List_of_industrial_disasters (last modified on January 15, 2016).

17. Braer, UK, 1993, retrieved on January 23, 2015 from website: http://www.itopf. com/in-action/case studies/case-study/braer-uk-1993/.

18. San Bruno Explosion, retrieved on September 12, 2010 from website: http://web. archive.org/web/20100913065813/http://www.huffington.post.com/2010/09/ san.bruno-explosion-resid-n-713330.html.

19. Albaiges, J., Vilas, F., Morales-Nin, B., The prestige: A scientific response, *Marine Pollution Bulletin*, 53(5–7), 2006, 15–20.

20. Guillen, A.V., Prestige and the law: Regulations and compensation, *Proceedings of the Annual Conference on Oil Pollution*, 2004, pp. 50–54.

21. Prestige, Spain/France, 2002, retrieved on January 23, 2015 from website: http:// www.itopf.com/in-action/case-studies/case-study/prestige-spainfrance-2002.

22. ASSE, *Dictionary of Terms Used in the Safety Profession*, 3rd Edition, American Society of Safety Engineers, Des Plaines, Illinois, 1988.

23. MIL-STD-721, *Definitions of Effectiveness Terms for Reliability, Maintainability, Human Factors, and Safety*, Department of Defense, Washington, DC.
24. Omdahl, T.P., Editor, *Reliability, Availability, and Maintainability (RAM) Dictionary*, ASQC Quality Press, Milwaukee, Wisconsin, 1988.
25. Dhillon, B.S., *Human Reliability: With Human Factors*, Pergamon Press, Inc., New York, 1986.
26. Neresky, J.J., Reliability definitions, *IEEE Transactions on Reliability*, 19, 1970, 198–200.

2

Basic Mathematical Concepts

2.1 Introduction

As in the development of other areas of engineering, mathematics has played an important role in the development of the safety and reliability fields in the oil and gas industry too. The history of mathematics may be traced back to the development of our currently used number symbols, often referred to as the "Hindu-Arabic numeral system" in the published literature [1]. Among the early evidence of the use of these numerals are the notches found on stone columns erected around 250 BC by the Scythian Emperor of India named Asoka [1].

The earliest reference to the probability concept may be traced back to a gambler's manual written by Girolamo (1501–1576), in which he considered some interesting issues on probability [1,2]. However, Pierre Fermat (1601–1665) and Blaise Pascal (1623–1662) were the first who solved independently and correctly the problem of dividing the winnings in a game of chance [1,2]. Boolean algebra, which plays a significant role in modern probability theory, is named after the mathematician George Boole (1815–1864), who published the pamphlet "The Mathematical Analysis of Logic: Being an Essay towards a Calculus of Deductive Reasoning" in 1847 [1–3].

Laplace transforms, often used in the area of reliability to find solutions to first-order differential equations, were developed by the French mathematician named Pierre-Simon Laplace (1749–1827). Additional information on the history of mathematics and probability is available in References 1 and 2. This chapter presents basic mathematical concepts considered useful to understand subsequent chapters of this book.

2.2 Boolean Algebra Laws

Boolean algebra is used to a degree in safety and reliability studies concerning oil and gas industry. Some of its laws that are considered useful to understand subsequent chapters of this book are presented below [3–6]:

Commutative law:

$$A \cdot B = B \cdot A \tag{2.1}$$

where
   A is an arbitrary set or event.
   B is an arbitrary set or event.
   Dot (.) denotes the intersection of sets. Sometimes Equation 2.1 is written without the dot (e.g., AB), but it still conveys the same meaning.

$$A + B = B + A \tag{2.2}$$

where
   + denotes the union of sets or events.

Idempotent law:

$$A + A = A \tag{2.3}$$

$$AA = A \tag{2.4}$$

Absorption law:

$$A(A + B) = A \tag{2.5}$$

$$A + (AB) = A \tag{2.6}$$

Associative law:

$$(A + B) + C = A + (B + C) \tag{2.7}$$

where
   C is an arbitrary set or event.

$$(AB)C = A(BC) \tag{2.8}$$

Distributive law:

$$(A+B)(A+C) = A + BC \qquad (2.9)$$

$$A(B+C) = AB + AC \qquad (2.10)$$

2.3 Probability Definition and Properties

The probability is defined as follows [7,8]:

$$P(A) = \lim_{n \to \infty} \left(\frac{N}{n} \right) \qquad (2.11)$$

where
$P(A)$ is the occurrence probability of event A.
N is the number of times event A occurs in the n repeated experiments.

Some of the basic properties of probability are as follows [5,7,8]:

- The properties of occurrence of event, say X, is

$$0 \le P(X) \le 1 \qquad (2.12)$$

- The probability of occurrence and nonoccurrence of event, say X, is always:

$$P(X) + P(\overline{X}) = 1 \qquad (2.13)$$

where
$P(X)$ is the probability of the occurrence of event X.
$P(\overline{X})$ is the probability of the nonoccurrence of event X.

- The probability of an intersection of n independent events is

$$P(X_1 X_2 X_3 \ldots X_n) = P(X_1)P(X_2)P(X_3) ----P(X_n) \qquad (2.14)$$

where
$P(X_i)$ is the probability of the occurrence of event X_i for $i = 1, 2, 3, \ldots, n$.

- The probability of the union of n mutually exclusive events is

$$P(X_1 + X_2 + X_3 + \cdots + X_n) = \sum_{i=1}^{n} P(X_i) \qquad (2.15)$$

- The probability of the union of n independent events is

$$P(X_1 + X_2 + X_3 + \cdots + X_n) = 1 - \prod_{i=1}^{n} (1 - P(X_i)) \qquad (2.16)$$

EXAMPLE 2.1

A system used in the oil and gas industry is composed of two critical subsystems X_1 and X_2. The failure of either subsystem can result in an accident. The probability of failure of subsystems X_1 and X_2 is 0.04 and 0.03, respectively.

Calculate the probability of the occurrence of an accident in the oil and gas industry system if both these subsystems fail independently. By substituting the given data values into Equation 2.16, we get

$$P(X_1 + X_2) = 1 - \prod_{i=1}^{2} (1 - P(X_i))$$

$$= P(X_1) + P(X_2) - P(X_1)P(X_2)$$

$$= 0.04 + 0.03 - (0.04)(0.03)$$

$$= 0.0688$$

Thus, the probability of the occurrence of an accident in the oil and gas industry system is 0.0688.

2.4 Useful Mathematical Definitions

This section presents a number of mathematical definitions that are considered to be useful in performing various types of safety and reliability studies concerned with systems used in the oil and gas industry.

2.4.1 Cumulative Distribution Function

For continuous random variables, this is defined by [7,8]

$$F(t) = \int_{-\infty}^{t} f(x)dx \tag{2.17}$$

where
    $F(t)$ is the cumulative distribution function.
    x is a continuous random variable.
    $f(x)$ is the probability density function.

For $t = \infty$, Equation 2.17 becomes

$$F(\infty) = \int_{-\infty}^{\infty} f(x)dx$$
$$= 1 \tag{2.18}$$

It means that the total area under the probability density curve is equal to unity.

Usually, in safety and reliability studies concerned with systems used in the oil and gas industry, Equation 2.17 is simply written as

$$F(t) = \int_{0}^{t} f(x)dx \tag{2.19}$$

EXAMPLE 2.2

Assume that the probability (i.e., failure) density function of a system used in the oil and gas industry is expressed by

$$f(t) = \alpha e^{-\alpha t}, \quad \text{for } t \geq 0, \ \alpha > 0 \tag{2.20}$$

where
    α is the failure rate of the system.
    t is a continuous random variable (i.e., time).
    $f(t)$ is the probability density function (generally, in the area of reliability engineering, it is referred to as the failure density function).

With the aid of Equation 2.20, obtain an expression for the oil and gas industry system cumulative distribution function.

By substituting Equation 2.20 into Equation 2.19, we obtain

$$F(t) = \int_{0}^{t} \alpha e^{-\alpha t} dt$$
$$= 1 - e^{-\alpha t} \tag{2.21}$$

Thus, Equation 2.21 is the expression for the oil and gas industry system cumulative distribution function.

2.4.2 Probability Density Function

For a continuous random variable, the probability density function is expressed by [7,9]

$$f(t) = \frac{dF(t)}{dt} \tag{2.22}$$

where
　　$F(t)$ is the cumulative distribution function.
　　$f(t)$ is the probability density function.

EXAMPLE 2.3

With the aid of Equation 2.21, prove that Equation 2.20 is the probability density function.

　　By inserting Equation 2.21 into Equation 2.22, we obtain

$$f(t) = \frac{d(1 - e^{-\alpha t})}{dt}$$

$$= \alpha e^{-\alpha t} \tag{2.23}$$

Equations 2.20 and 2.23 are identical.

2.4.3 Expected Value

The expected value of a continuous random variable is defined by [7]

$$E(t) = \mu = \int_{-\infty}^{\infty} t f(t)\,dt \tag{2.24}$$

where
　　$E(t)$ is the expected value of the continuous random variable t.
　　μ is the mean value.

EXAMPLE 2.4

Find the mean value (i.e., expected value) of the probability (failure) density function expressed by Equation 2.20.

　　By inserting Equation 2.20 into Equation 2.24, we obtain

$$E(t) = \int_{0}^{\infty} t \alpha e^{-\alpha t}\,dt$$

$$= \left[-t e^{-\alpha t} \right]_{0}^{\infty} - \left[-\frac{e^{-\alpha t}}{\alpha} \right]_{0}^{\infty} \tag{2.25}$$

$$= \frac{1}{\alpha}$$

Thus, the mean value (i.e., expected value) of the probability (failure) density function expressed by Equation 2.20 is given by Equation 2.25.

2.4.4 Laplace Transform Definition

The Laplace transform of a function, say $f(t)$, is defined by [1,10,11]

$$f(s) = \int_0^\infty f(t)e^{-st}\, dt \qquad (2.26)$$

where
   s is the Laplace transform variable.
   $f(s)$ is the Laplace transform of function $f(t)$.
   t is a variable.

EXAMPLE 2.5

Obtain the Laplace transform of the following function:

$$f(t) = e^{-\alpha t} \qquad (2.27)$$

where
   α is a constant.

By inserting Equation 2.27 into Equation 2.26, we get

$$f(s) = \int_0^\infty e^{-\alpha t}e^{-st}\, dt$$

$$= \left. \frac{e^{-(s+\alpha)t}}{(s+\alpha)} \right|_0^\infty \qquad (2.28)$$

$$= \frac{1}{s+\alpha}$$

Thus, Equation 2.28 is the Laplace transform of Equation 2.27.

EXAMPLE 2.6

Obtain the Laplace transform of the following function:

$$f(t) = 1 \qquad (2.29)$$

By substituting Equation 2.29 into Equation 2.26, we obtain

TABLE 2.1

Laplace Transforms of Some Functions

$f(t)$	$f(s)$
C (a constant)	C/s
t^m, for $m = 0, 1, 2, 3,...$	$m!/s^{m}+1$
$e^{-\alpha t}$	$1/(s + \alpha)$
$tf(t)$	$-df(s)/ds$
$df(t)/dt$	$s\,f(s) - f(0)$
$\alpha_1 f_1(t) + \alpha_2 f_2(t)$	$\alpha_1 f_1(s) + \alpha_2 f_2(s)$
t	$1/s^2$
$te^{-\alpha t}$	$1/(s + \alpha)^2$

$$f(s) = \int_0^\infty 1e^{-st}dt$$

$$= \frac{e^{-st}}{-s}\bigg|_0^\infty \tag{2.30}$$

$$= \frac{1}{s}$$

Thus, Equation 2.30 is the Laplace transform of Equation 2.29.

Laplace transforms of some commonly occurring functions in oil and gas industry system safety and reliability-related analysis studies are presented in Table 2.1 [10–12].

2.4.5 Laplace Transform: Final-Value Theorem

If the following limits exist, then the final-value theorem may be expressed as [8,10,11]

$$\lim_{t \to \infty} f(t) = \lim_{s \to 0} \left[sf(s) \right] \tag{2.31}$$

EXAMPLE 2.7

Prove, by using the following equation, that the right-hand side of Equation 2.31 is equal to its left-hand side:

$$f(t) = \frac{\alpha_1}{(\alpha_1 + \alpha_2)} + \frac{\alpha_2}{(\alpha_1 + \alpha_2)} e^{-(\alpha_1 + \alpha_2)t} \tag{2.32}$$

where
  α_1 and α_2 are the constants.

With the aid of Table 2.1, we get the following Laplace transforms of Equation 2.32:

$$f(s) = \frac{\alpha_1}{s(\alpha_1 + \alpha_2)} + \frac{\alpha_2}{(\alpha_1 + \alpha_2)} \frac{1}{(s + \alpha_1 + \alpha_2)} \tag{2.33}$$

By substituting Equation 2.33 into the right-hand side of Equation 2.31, we get

$$\lim_{s \to 0} s \left[\frac{\alpha_1}{s(\alpha_1 + \alpha_2)} + \frac{\alpha_2}{(\alpha_1 + \alpha_2)} \frac{1}{(s + \alpha_1 + \alpha_2)} \right] = \frac{\alpha_1}{\alpha_1 + \alpha_2} \tag{2.34}$$

By inserting Equation 2.32 into the left-hand side of Equation 2.31, we obtain

$$\lim_{t \to \infty} \left[\frac{\alpha_1}{(\alpha_1 + \alpha_2)} + \frac{\alpha_2}{(\alpha_1 + \alpha_2)} e^{-(\alpha_1 + \alpha_2)t} \right] = \frac{\alpha_1}{\alpha_1 + \alpha_2} \tag{2.35}$$

As the right-hand sides of Equations 2.34 and 2.35 are identical, it proves that the right-hand side of Equation 2.31 is equal to its left-hand side.

2.5 Probability Distributions

Although there are a large number of probability/statistical distributions in the published literature, this section presents just five such distributions considered useful for performing oil and gas industry system safety and reliability-related studies [13–15].

2.5.1 Binomial Distribution

This distribution has applications in many combinational-type safety and reliability-related problems, and sometimes it is also called a Bernoulli (after its founder Jakob Bernoulli [1654–1705] distribution [1]).

This probability distribution becomes very helpful in situations where one is concerned with the probabilities of outcome such as the number of failures in a sequence of n trials. It is to be noted that for binomial distribution, each trial has two possible outcomes (e.g., success and failure), and the probability of each trial remains unchanged or constant.

The distribution probability density function, $f(x)$, is expressed by [8,9,13]

$$f(x) = \binom{n}{j} p^x q^{n-x}, \text{ for } x = 0, 1, 2, \ldots, n \tag{2.36}$$

where
  x is the number of failures in n trials.
  q is the single trial probability of failure.
  p is the single trial probability of success.

$$\binom{n}{j} = \frac{n!}{(n-j)! j!}$$

The cumulative distribution function is expressed by

$$F(x) = \sum_{j=0}^{x} \binom{n}{j} p^j q^{n-j} \tag{2.37}$$

where
  $F(x)$ is the cumulative distribution function or the probability of x or less failures in n trials.

2.5.2 Exponential Distribution

This is one of the simplest continuous random variable distribution frequently used in the industry, particularly in performing reliability studies because many engineering items exhibit a constant hazard rate during their useful life [14]. In addition, it is relatively easy to handle in performing reliability analysis-related studies.

The distribution probability density function is expressed by [8,14]

$$f(t) = \alpha e^{-\alpha t}, \quad \text{for } \alpha > 0, \ t \geq 0 \tag{2.38}$$

where
  α is the distribution parameter.
  t is the time (i.e., a continuous random variable).
  $f(t)$ is the probability density function.

By inserting Equation 2.38 into Equation 2.19, we get the following expression for the cumulative distribution function:

$$F(t) = 1 - e^{-\alpha t} \tag{2.39}$$

With the aid of Equations 2.38 and 2.24, we obtain the following expression for the distribution mean value (i.e., expected value):

$$\mu = E(t) = \frac{1}{\alpha} \qquad (2.40)$$

EXAMPLE 2.8

Assume that the mean time to failure of a system used in the oil and gas industry is 1500 h. Calculate the probability of failure of the system during a 700-h mission with the aid of Equations 2.40 and 2.39.

By inserting the specified data value into Equation 2.40, we obtain

$$\alpha = \frac{1}{1500} = 0.00067 \text{ failures per hour}$$

By substituting the calculated and the specified data values into Equation 2.39, we get

$$F(700) = 1 - e^{-(0.00067)(700)}$$

$$= 0.3729$$

Thus, the probability of failure of the oil and gas industry system during the 700-h mission is 0.3729.

2.5.3 Rayleigh Distribution

This continuous random variable probability distribution is named after its founder, John Rayleigh (1842–1919), and its probability density function is defined by [1,8]

$$f(t) = \left(\frac{1}{\theta^2}\right) t e^{-(t/\theta)^2}, \quad \text{for } \theta > 0, \ t > 0 \qquad (2.41)$$

where
θ is the distribution parameter.

By substituting Equation 2.41 into Equation 2.19, we obtain the following expression for the cumulative distribution function:

$$F(t) = 1 - e^{-(t/\theta)^2} \qquad (2.42)$$

By inserting Equation 2.41 into Equation 2.24, we get the following equation for the distribution mean value (i.e., expected value):

$$\mu = E(t) = \theta\Gamma\left(\frac{3}{2}\right) \tag{2.43}$$

where
$\Gamma(.)$ is the gamma function and is defined by

$$\Gamma(n) = \int_0^\infty t^{n-1}e^{-t}dt, \quad \text{for } n > 0 \tag{2.44}$$

2.5.4 Weibull Distribution

This continuous random variable probability distribution was developed in the early 1950s by Walliodi Weibull, a Swedish professor in mechanical engineering [15]. The probability density function for the distribution is defined by

$$f(t) = \frac{at^{a-1}}{\alpha^a}e^{-(t/\alpha)^a}, \quad \text{for } \alpha > 0, \ a > 0, \ t \geq 0 \tag{2.45}$$

where
α and a are the distribution scale and shape parameters, respectively.

By inserting Equation 2.45 into Equation 2.19, we obtain the following equation for the cumulative distribution function:

$$F(t) = 1 - e^{-(t/\alpha)^a} \tag{2.46}$$

Using Equations 2.45 and 2.24, we obtain the following equation for the distribution mean value (i.e., expected value):

$$\mu = E(t) = \alpha\Gamma\left(1 + \frac{1}{a}\right) \tag{2.47}$$

For $a = 1$ and $a = 2$, the exponential and Rayleigh distributions are the special cases of this distribution, respectively.

2.5.5 Bathtub Hazard Rate Curve Distribution

This continuous random variable probability distribution was developed in 1981 [16]. In the published literature by other authors around the world, it is

generally referred to as *Dhillon distribution/law/model* [17–37]. The distribution can represent bathtub-shape, decreasing and increasing hazard rates.
   The distribution probability density function is defined by [16]

$$f(t) = a\alpha(\alpha t)^{a-1} e^{-\left\{ e^{(\alpha t)^a} - (\alpha t)^a - 1 \right\}}, \quad \text{for } a > 0,\ \alpha > 0,\ t \geq 0 \tag{2.48}$$

where
   α and a are the distribution scale and shape parameters, respectively.

By substituting Equation 2.48 into Equation 2.19, we obtain the following equation for the cumulative distribution function:

$$F(t) = 1 - e^{-\left\{ e^{(\alpha t)^a} - 1 \right\}} \tag{2.49}$$

For $a = 0.5$, this probability distribution gives the bathtub-shape hazard rate curve, and for $a = 1$, it gives the extreme value probability distribution. In other words, the extreme value probability distribution is the special case of this probability distribution at $a = 1$.

2.6 Solving First-Order Linear Differential Equations Using Laplace Transforms

Usually, Laplace transforms are used in finding solutions to first-order differential equations in reliability and safety analysis-related studies of systems used in the oil and gas industry.
   The example presented below demonstrates the finding of solutions to a set of linear first-order differential equations, describing an oil and gas industry system with respect to safety and reliability, using Laplace transforms.

EXAMPLE 2.9

Assume that a system used in the oil and gas industry can be in any of the three states: operating normally, failed safely, and failed unsafely. The following three first-order linear differential equations describe the oil and gas industry system under consideration:

$$\frac{dP_0(t)}{dt} + (\lambda_s + \lambda_u)P_0(t) = 0 \tag{2.50}$$

$$\frac{dP_s(t)}{dt} - \lambda_s P_0(t) = 0 \tag{2.51}$$

$$\frac{dP_u(t)}{dt} - \lambda_u P_0(t) = 0 \tag{2.52}$$

where

$P_i(t)$ is the probability that the oil and gas industry system is in state i at time t, for $i = 0$ (operating normally), $i = s$ (failed safely), and $i = u$ (failed unsafely).

λ_s is the oil and gas industry system failing safely constant failure rate.

λ_u is the oil and gas industry system failing unsafely constant failure rate.

At time $t = 0$, $P_0(0) = 1$, $P_s(0) = 0$, and $P_u(0) = 0$.

Solve differential Equations 2.50 through 2.52 using Laplace transforms.

With the aid of Table 2.1, differential Equations 2.50 through 2.52, and the specified initial conditions, we obtain

$$sP_0(s) - 1 + (\lambda_s + \lambda_u)P_0(s) = 0 \tag{2.53}$$

$$sP_s(s) - \lambda_s P_0(s) = 0 \tag{2.54}$$

$$sP_u(s) - \lambda_u P_0(s) = 0 \tag{2.55}$$

By solving Equations 2.53 through 2.55, we get

$$P_0(s) = \frac{1}{(s + \lambda_s + \lambda_u)} \tag{2.56}$$

$$P_s(s) = \frac{\lambda_s}{s(s + \lambda_s + \lambda_u)} \tag{2.57}$$

$$P_u(s) = \frac{\lambda_u}{s(s + \lambda_s + \lambda_u)} \tag{2.58}$$

By taking the inverse Laplace transforms of Equations 2.56 through 2.58, we obtain:

$$P_0(t) = e^{-(\lambda_s + \lambda_u)t} \tag{2.59}$$

$$P_s(t) = \frac{\lambda_s}{(\lambda_s + \lambda_u)}\left[1 - e^{-(\lambda_s + \lambda_u)t}\right] \tag{2.60}$$

$$P_u(t) = \frac{\lambda_u}{\lambda_s + \lambda_u}\left[1 - e^{-(\lambda_s + \lambda_u)t}\right] \tag{2.61}$$

Thus, Equations 2.59 through 2.61 are the solutions to linear differential Equations 2.50 through 2.52.

PROBLEMS

1. Prove Equation 2.9.
2. Write an essay on the history of mathematics, including probability theory.
3. Mathematically define probability.
4. What is idempotent law?
5. What are the basic properties of probability?
6. Define the following items for continuous random variables:
   - Probability density function
   - Cumulative distribution function
   - Expected value
7. Write down probability density function for the bathtub hazard rate curve distribution.
8. What are the special case probability distributions of the Weibull distribution?
9. Prove Equations 2.56 through 2.58 with the aid of Equations 2.53 through 2.55. What is the sum of Equations 2.56 through 2.58?
10. What is the special case distribution of the bathtub hazard rate curve distribution?

References

1. Eves, H., *An Introduction to the History of Mathematics*, Holt, Rinehart, and Winston, New York, 1976.
2. Owen, D.B., Editor, *On the History of Statistics and Probability*, Marcel Dekker, New York, 1976.
3. Lipschutz, S., *Set Theory*, McGraw-Hill Book Company, New York, 1964.
4. Speigel, M.R., *Statistics*, McGraw-Hill Book Company, New York, 1961.
5. Lipschutz, S., *Probability*, McGraw-Hill Book Company, New York, 1965.
6. *Fault Tree Handbook*, Report No. NUREG-0492, U.S. Nuclear Regulatory Commission, Washington, DC, 1981.
7. Mann, N.R., Schafer, R.E., Singpurwalla, N.P., *Methods for Statistical Analysis of Reliability and Life Data*, John Wiley and Sons, New York, 1974.
8. Dhillon, B.S., *Design Reliability: Fundamentals and Applications*, CRC Press, Boca Raton, Florida, 1999.
9. Shooman, M.L., *Probabilistic Reliability: An Engineering Approach*, McGraw-Hill Book Company, New York, 1968.

10. Spiegel, M.R., *Laplace Transforms*, McGraw-Hill Book Company, New York, 1965.
11. Oberhettinger, F., Badii, L., *Tables of Laplace Transforms*, Springer-Verlag, New York, 1973.
12. Nixon, F.E., *Handbook of Laplace Transformation: Fundamentals, Applications, Tables, and Examples*, Prentice-Hall, Inc., Englewood Cliffs, New Jersey, 1960.
13. Patel, J.K., Kapadia, C.H., Owen, D.H., *Handbook of Statistical Distributions*, Marcel Dekker, New York, 1976.
14. Davis, D.J., An analysis of some failure data, *Journal of the American Statistical Association*, 47, 1952, 113–150.
15. Weibull, W., A statistical distribution of wide applicability, *Journal of Applied Mechanics*, 18, 1951, 293–297.
16. Dhillon, B.S., Life distributions, *IEEE Transactions on Reliability*, 30, 1981, 457–460.
17. Henze, N., Meintnis, S.G., Recent and classical tests for exponentially: A partial review with comparisons, *Metrica*, 61, 2005, 29–45.
18. Baker, R.D., Non-parametric estimation of the renewal function, *Computers Operations Research*, 20(2), 1993, 167–178.
19. Hollander, M., Laird, G., Song, K.S., Non-parametric interference for the proportionality function in the random censorship model, *Journal of Nonparametric Statistics*, 15(2), 2003, 151–169.
20. Nimoto, N., Zitikis, R., The Atkinson index, the Moran statistic, and testing exponentiality, *Journal of the Japan Statistical Society*, 38(2), 2008, 187–205.
21. Kunitz, H., A new class of bathtub-shaped hazard rates and its application in comparison of two test-statistics, *IEEE Transactions on Reliability*, 38(3), 1989, 351–354.
22. Cabana, A., Cabana, E.M., Goodness-of-fit to the exponential distribution, focused on Weibull alternatives, *Communications in Statistics-Simulation and Computation*, 34, 2005, 711–723.
23. Grane, A., Fortiana, J., A directional test of exponentiality based on maximum correlations, *Metrika*, 73, 2011, 255–274.
24. Jammalamadaka, S.R., Taufer, E., Testing exponentiality by comparing the empirical distribution function of the normalized spacings with that of the original data, *Journal of Nonparametric Statistics*, 15(6), 2003, 719–729.
25. Morris, K., Szynal, D., Goodness-of-fit tests based on characterizations involving moments or order statistics, *International Journal of Pure and Applied Mathematics*, 38(1), 2007, 83–121.
26. Jammalamadaka, S.R., Taufer, E., Use of mean residual life in testing departures from exponentiality, *Journal of Nonparametric Statistics*, 18(3), 2006, 277–292.
27. Kunitz, H. Pamme, H., The mixed gamma aging model in life data analysis, *Statistical Papers*, 34, 1993, 303–318.
28. Meintanis, S.G., A class of tests for exponentiality based on a continuum of moment conditions, *Kybernetika*, 45(6), 2009, 946–959.
29. Na, M.H., Spline hazard rate estimation using censored data, *Journal of KSIAM*, 3(2), 1999, 99–106.
30. Morris, K., Szynal, D., Some U-statistics in goodness-of-fit tests derived from characterizations via record values, *International Journal of Pure and Applied Mathematics*, 46(4), 2008, 339–414.
31. Nam, K.H., Park, D.H., Failure rate for Dhillon model, *Proceedings of the Spring Conference of the Korean Statistical Society*, 1997, pp. 114–118.

32. Szynal, D., *Goodness-of-Fit Tests Derived from Characterizations of Continuous Distributions, Stability in Probability*, Banach Center Publications, Vol. 90, Institute of Mathematics, Polish Academy of Sciences, Warszawa, Poland, 2010, pp. 203–223.
33. Nam, K.H., Park, D.H., A study on trend changes for certain parametric families, *Journal of the Korean Society for Quality Management*, 23(3), 1995, 93–101.
34. Szynal, D., Wolynski, W., Goodness-of-fit tests for exponentiality and Rayleigh distribution, international, *Journal of Pure and Applied Mathematics*, 78(5), 2012, 751–772.
35. Noughabi, H.A., Arghami, N.R., Testing exponentiality based on characterizations of the exponential distribution, *Journal of Statistical Computation and Simulation*, 1(1), 2011, 1–11.
36. Nam, K.H., Chang, S.J., Approximation of the renewal function for Hjorth model and Dhillon model, *Journal of the Korean Society for Quality Management*, 34(1), 2006, 34–39.
37. Srivastava, S.K., Validation analysis of Dhillon model on different real data sets for reliability modeling, *International Journal of Advanced Foundation and Research in Computer*, 1(9), 2014, 18–31.

3

Safety and Reliability Basics

3.1 Introduction

The history of safety in the modern times may be traced back to 1868, when a patent for a barrier safeguard was awarded in the United States [1]. In 1893, the U.S. Congress passed the Railway Safety Act and, in 1912, the cooperative Safety Congress met for the first time in Milwaukee, Wisconsin [1–3]. Today, the field of safety has developed into many areas, including workplace safety, patient safety, robot safety, and software safety.

The history of the reliability field goes back to the early 1930s when probability concepts were applied to problems associated with electric power generation [4,5]. During World War II, German scientists applied the basic reliability concepts for improving reliability of their V1 and V2 rockets [6]. Today, the field of reliability has become a well-developed discipline and has branched out into many specialized areas, including power system reliability, software reliability, human reliability, and mechanical reliability.

Additional information on the history of both safety and reliability fields is available in References 3 and 6.

This chapter presents various safety and reliability fields' basics considered useful to understand subsequent chapters of this book.

3.2 Need for Safety and Safety and Engineers

The desire to be safe and secure has always been an important issue to humans. For example, early humans took appropriate measures to guard against natural hazards surrounding them. Moreover, in 2000 BC, Hammurabi, an ancient Babylonian ruler, developed a code referred to as Code of Hammurabi. This code included clauses on a number of items, including monetary damages against individuals who caused injury to others and allowable fees for physicians of that time [1,7,8].

Today, safety has become an important issue around the globe because each year, a very large number of people get seriously injured or die due to workplace-related and other accidents. For example, in 1996 in the United States alone, according to the findings of the National Safety Council, there were 93,400 deaths and a very large number of disabling injuries due to accidents [9]. The overall cost of these accidents was estimated to be about $121 billion. There are many other factors that also play an important role in demanding the need for better safety, including public pressure, increasing number of law suits, and government regulations.

Nowadays, engineering products have become highly sophisticated and complex. Safety of these products has become a challenging issue to engineers. Owing to global competition and other factors, engineers are pressured to complete new designs rapidly and at lower costs. Past experiences clearly indicate that this, in turn, generally leads to greater design-related deficiencies, errors, and causes for the occurrence of accidents. Design-related deficiencies can cause or, directly or indirectly, contribute to accidents. The design deficiency may result because a design/designer [3,8]

- Relies on the users of the product to avoid an accident
- Overlooked eliminating or reducing the human error occurrence
- Is confusing, incorrect, or unfinished
- Overlooked foreseeing an item's unexpected application or its potential consequences
- Does not properly consider or determine the consequences of failure, error, action, or omission
- Overlooked giving a warning properly of a potential hazard
- Violates potential users' normal tendencies/capabilities
- Overlooked prescribing an effective operational procedure in situations where a hazard might exist
- Creates an unsafe characteristic of an item
- Incorporates poor warning mechanisms to eradicate hazards rather than providing a safe design
- Overlooked providing satisfactory protection in personal protective equipment of users/workers
- Creates an arrangement of operating controls and other devices that considerably increase reaction time in emergency situations or is quite conducive to errors
- Places unreasonable stress on operators

3.3 Safety Management Principles

There are many safety management principles. Ten of these principles are as follows [8,10]:

1. The safety system should be tailored to fit the culture of company/organization.
2. The key to successful line safety performance is management procedures that clearly factor in accountability effectively.
3. The function of safety is to discover and define operational errors that cause accidents.
4. Causes that result in unsafe behavior can be controlled, identified, and classified.
5. Management should manage safety just like managing other functions in the organization, more clearly, by planning, organizing, controlling, and setting attainable goals.
6. Management has the responsibility for making changes to the environment that lead to unsafe behavior because unsafe behavior is the result of normal individuals reacting to their surrounding environment.
7. Symptoms that indicate something is not right in the safety management are an accident, an unsafe act, and an unsafe condition.
8. The key for having a good safety system is clearly visible top management support, flexibility, worker participation, and so on.
9. Circumstances that can be predicted to lead to serious injuries are high-energy sources, nonroutine activities, nonproductive activities, and certain construction situations.
10. In developing a good safety system, consider with care three major subsystems: the managerial, the behavioral, and the physical.

3.4 Product Hazard Classifications and Product Safety Organization Tasks

There are many product-related hazards. They may be categorized into six classifications as shown in Figure 3.1 [8,11].

The energy hazards may be grouped into two categories: potential energy and kinetic energy. The potential energy hazards pertain to items that store energy. These items include springs, electronic capacitors, compressed gas

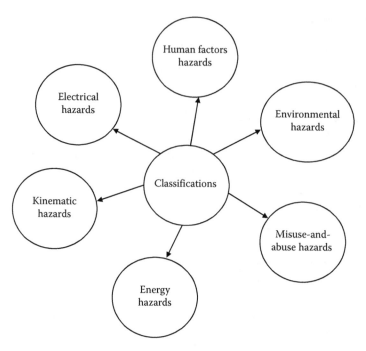

FIGURE 3.1
Product hazard classifications.

receivers, and counterbalancing weights. During the equipment servicing process, such hazards are very important because stored energy can cause serious injury when released suddenly. The kinetic energy hazards pertain to items that have energy because of their motion. Three examples of such items are fan blades, flywheels, and loom shuttles. Any object that interferes with the motion of such items can experience substantial damage.

The human factors hazards are associated with poor design with respect to humans, more specifically, to their physical strength, intelligence, weight, education, visual angle, computational ability, visual acuity, height, length of reach, and so on.

The environmental hazards may be categorized into two groups: external and internal. The external hazards are the hazards posed by the product under consideration during its entire life span. These hazards include service-life operation hazards, maintenance hazards, and disposal hazards. The internal hazards are associated with the changes in the surrounding environment that result in internally damaged product. The hazards such as these can be minimized or eliminated by considering with care factors such as extremes of temperatures, vibrations, ambient noise levels, atmospheric contaminants, illumination level, and electromagnetic radiation during the design phase.

The electrical hazards have two main components: shock hazard and electrocution hazard. The major electrical hazard to product/property stems from electrical faults, often referred to as short circuits.

Misuse-and-abuse hazards are associated with the product usage by humans. Past experiences clearly indicate that product misuse can lead to serious injuries. Product abuse can also lead to hazardous injuries/situations, and some of the causes for the abuse are lack of adequate maintenance and poor operating practices.

Finally, the kinematic hazards are concerned with situations where parts/components come together while moving and lead to possible crushing, pinching, or cutting of any object/item caught between them.

An organization involved with product safety performs many safety-related tasks. Thirteen of these tasks are as follows [12,13]:

1. Develop appropriate programs and directives concerned with product safety
2. Develop mechanisms by which the safety program can be monitored effectively
3. Review safety-related field reports and customer complaints
4. Review nongovernment and government requirements concerned with product safety
5. Review the product to establish whether all the potential hazards have been controlled or eradicated altogether
6. Review all the proposed product operation and maintenance documents with respect to safety
7. Take part in reviewing accident-related claims or recall actions by government agencies and recommend remedial measures for justifiable claims or recalls
8. Review all product test reports for determining shortcomings or trends with respect to safety
9. Develop safety criteria on the basis of applicable governmental and voluntary standards for use by company design professionals, subcontractors, and vendors
10. Review hazards and mishaps in existing similar products for avoiding repetition of such hazards in new products
11. Review all warning labels with respect to safety factors such as satisfying requirements of, adequacy, and compatibility to warnings in the instruction manuals that are to be placed on products
12. Provide appropriate assistance to designers in selecting alternative means for eradicating or controlling hazards or other safety-associated problems in preliminary designs

13. Determine if items such as protective equipment, monitoring and warning devices, and emergency equipment are needed to handle or use the product

3.5 Common Causes of Work Injuries and Mechanical Injuries

Over the years, professionals working in the area of accident investigations have identified many different causes of work injuries. The common ones are shown in Figure 3.2 [2,14].

It is to be noted that according to a study conducted by the National Safety Council of the United States, about 31% of all work injuries are caused by overexertion [1].

In the industrial sector, including oil and gas, humans interact with various types of equipment to carry out tasks such as drilling, cutting, shaping, chipping, punching, stamping, abrading, and stitching. There are various types of injuries that can occur in performing tasks such as these. The six common ones are as follows [1,8,14]:

1. *Straining and spraining injuries*: These injuries are generally associated with the use of machines and other tasks. Examples of such injuries include spraining of ligaments or straining of muscles.

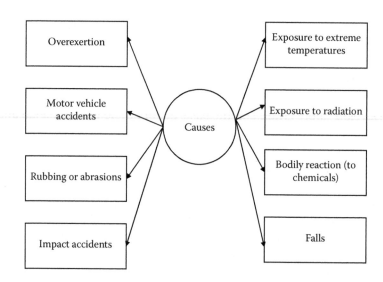

FIGURE 3.2
Common work injury causes.

2. *Cutting and tearing injuries*: These injuries occur when a body part of a person comes in contact with a sharp edge. The degree of severity of a cut or a tear depends greatly on the degree of damage to items such as skin, veins, muscles, and arteries.

3. *Crushing injuries*: These injuries take place when a body part is caught between two hard surfaces moving progressively together and crushing any object/item that comes in between.

4. *Breaking injuries*: These injuries are generally associated with machines used for deforming various types of engineering materials. A break in a bone is referred to as a fracture. In turn, fracture is classified into categories such as incomplete fracture, simple fracture, compound fracture, transverse fracture, complete fracture, oblique fracture, and comminuted fracture.

5. *Puncturing injuries*: These injuries occur when an object penetrates straight into the body of a person and pulls straight out. In the industrial setting, normally, these types of injuries pertain to punching machines because they have sharp tools.

6. *Shearing injuries*: These injuries pertain to shearing processes. In the area of manufacturing, power-driven shears are often used to perform tasks such as severing metal, paper, plastic, and elastomers. In the past, while using machines such as these, tragedies such as amputation of hands/fingers have occurred.

3.6 Accident Causation Theories

There are many accident causation theories [1]. Two of these theories are presented below.

3.6.1 Human Factors Accident Causation Theory

The basis for this theory is the assumption that accidents occur due to a chain of events directly or indirectly due to human error. It consists of three main factors, shown in Figure 3.3, that lead to the human error occurrence [1,14,15]. The factors are inappropriate activities, overload, and inappropriate response/incompatibility.

Inappropriate activities performed by an individual can be due to human error. For example, a person misjudged the degree of risk involved in a stated task and then performed the task on that misjudgment.

Overload is concerned with an imbalance between an individual's capacity at any time and the load that he/she is carrying in a given state. The capacity of a person is the product of factors such as physical condition,

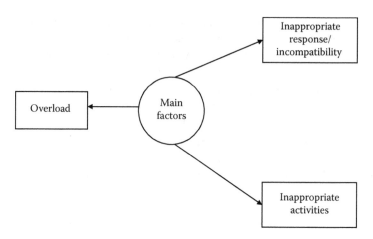

FIGURE 3.3
Main factors that lead to the human error occurrence.

fatigue, state of mind, stress, natural ability, and degree of training. The load carried by a person is composed of tasks for which he/she has responsibility along with additional burdens resulting from the environmental factors (i.e., distractions, noise, etc.), situational factors (i.e., level of risk, unclear instructions, etc.), and internal factors (i.e., personal problems, worry, emotional stress, etc.).

Inappropriate response/incompatibility is another major factor that can cause human errors. Three typical examples of inappropriate response are as follows:

1. A person detects a hazardous condition but takes no corrective action.
2. A person removes a safeguard from a machine to increase output.
3. A person disregards the specified safety procedures.

3.6.2 Domino Accident Causation Theory

This theory is encapsulated in 10 statements by H.W. Heinrich, known as the "Axioms of Industrial Safety" [16]. These 10 axioms/statements are as follows [14,16]:

- *Axiom I*: An accident can occur only when a person commits an unsafe act and/or there is a mechanical or physical hazard.
- *Axiom II*: Most accidents are the result of unsafe acts of humans.
- *Axiom III*: An unsafe act by a person or an unsafe condition does not always immediately lead to an injury/accident.

- *Axiom IV*: Supervisors play a key role in industrial accident prevention.
- *Axiom V*: The reasons why humans commit unsafe acts can be quite useful in selecting appropriate corrective measures.
- *Axiom VI*: The most useful accident prevention approaches are analogous to the productivity and quality methods.
- *Axiom VII*: There are two types of costs of an accident: direct and indirect. Some examples of the direct costs are liability claims, medical costs, and compensation.
- *Axiom VIII*: The severity of an injury is largely fortuitous and the specific accident that caused it is usually preventable.
- *Axiom IX*: Management should assume full safety responsibility with vigor because it is in the best position for achieving end results effectively.
- *Axiom X:* The occurrence of injuries results from a completed sequence of factors, the last/final one of which is the accident itself.

Furthermore, H.W. Heinrich also believed that there are the following five factors in the sequence of events leading up to an accident [1,2,14]:

1. *Ancestry and social environment*: In this factor, it is assumed that negative character traits such as recklessness, stubbornness, and avariciousness that might lead people to behave in an unsafe manner can be inherited through one's ancestry or acquired as a result of one's social surroundings or environment.
2. *Fault of person*: In this factor, it is assumed that negative character traits (whether inherited or acquired) such as recklessness, violent temper, nervousness, and ignorance of safety practices constitute proximate reasons for committing unsafe acts or for the existence of physical or mechanical-related hazards.
3. *Unsafe act/physical or mechanical hazard*: In this factor, it is assumed that unsafe acts by people (e.g., starting equipment/machinery without warning, removing safeguards, standing under suspended loads) and mechanical or physical hazards (e.g., unguarded gears, inadequate light, absence of guardrails, unguarded point of operation) are the direct causes for accident occurrences.
4. *Accident*: In this factor, it is assumed that events such as falls of people and striking of people by flying objects are the typical examples of accidents that cause injury.
5. *Injury*: In this factor, it is assumed that the typical injuries directly resulting from the accident occurrences include fractures and lacerations.

3.7 Occupational Stressors and Human Error Occurrence Reasons

There are many occupational stressors that may compromise safety and reliability in the oil and gas industry. These stressors may be grouped into four classifications as follows [2,14,16]:

1. *Workload-related stressors:* These stressors are concerned with work overload or work underload. In case of work overload, the job requirements exceed the ability of a person to satisfy them properly. Similarly, in case of work underload, the current duties being carried out by the person do not provide sufficient level of stimulation. Some examples of work underload are lack of opportunity for using acquired skills and expertise of the person, task repetitiveness, and the lack of any intellectual input.

2. *Occupational change-related stressors:* These stressors are concerned with factors that disrupt the proper functioning of physiological, behavioral, and cognitive patterns of the person.

3. *Occupational frustration-related stressors:* These stressors are concerned with the problems pertaining to occupational frustration. Three examples of these problems are the lack of effective communication, poor career development guidance, and the ambiguity of one's role.

4. *Miscellaneous stressors:* These stressors include all those stressors that are not included in the above three classifications. Some examples of the miscellaneous stressors are poor interpersonal relationships, too much noise, and too little or too much lighting.

There are many reasons for the occurrence of human errors. Some of the important ones are shown in Figure 3.4 [2,14,17].

3.8 Consequences of Human Error and Human Error Classifications

There is a wide range of consequences of human error and it can vary from minor consequences to very severe ones, for example, from insignificant delays in, say, oil and gas industry system performance to a very high loss of lives. Furthermore, consequences can significantly vary from one equipment/system to another, from one situation to another, and from one task to another.

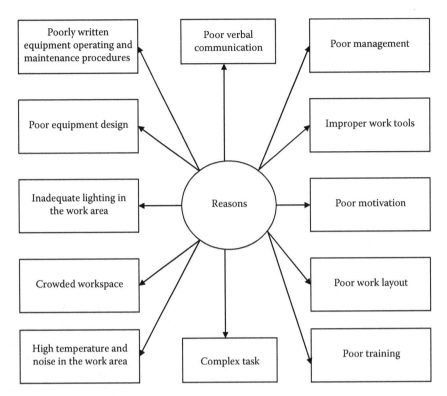

FIGURE 3.4
Important reasons for the occurrence of human errors.

In particular, with respect to equipment/system, human error consequences may be categorized under the following three classifications [2,14]:

1. *Classification I*: Operation of equipment/system is delayed quite
   significantly but not totally stopped.
2. *Classification II*: Operation of equipment/system is stopped
   completely.
3. *Classification III*: Delay in the operation of the equipment/system is
   insignificant.

In the area of engineering, various types of human errors occur. They may
be grouped under a number of classifications. The seven commonly used
classifications are as follows [2,17–20]:

1. *Classification I*: Design errors
2. *Classification II*: Assembly errors
3. *Classification III*: Operator errors

4. *Classification IV*: Maintenance errors
5. *Classification V*: Inspection errors
6. *Classification VI*: Handling errors
7. *Classification VII*: Contributory errors

Additional information on the above classification of human errors is available in References 6 and 17 through 21.

3.9 Bathtub Hazard Rate Curve

Generally, bathtub hazard rate curve is used to describe the failure rate of engineering items/systems and is shown in Figure 3.5.

The curve is known as the bathtub hazard rate curve because it resembles the shape of a bathtub. As shown in Figure 3.5, the curve is divided into three sections: burn-in period, useful-life period, and wear-out period. During the burn-in period, the item/system hazard rate decreases with time t; some of the reasons for the occurrence of failures during this period are poor quality control, poor manufacturing methods and processes, inadequate debugging, human error, and substandard materials and workmanship [6,22]. Other terms used in the published literature for this decreasing hazard rate region are "debugging region," "infant mortality region," and "break-in region."

During the useful-life period, the hazard rate remains constant. Some of the reasons for the occurrence of failures in this region are higher random stress than expected, low safety factors, undetectable defects, human errors, abuse, and natural failures.

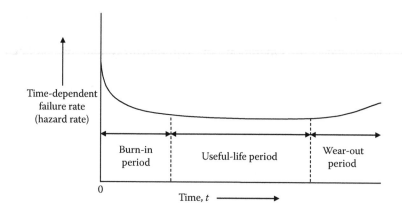

FIGURE 3.5
Bathtub hazard rate curve.

Finally, during the wear-out period, the hazard rate increases with time t due to reasons such as follows [6,22]:

- Wear due to friction, corrosion, and creep
- Wear due to aging
- Poor maintenance
- Wrong overhaul practices
- Short designed-in life of the item/system under consideration

The following equation obtained from a probability distribution presented in Chapter 2 can be used to represent the bathtub hazard rate curve, shown in Figure 3.5, mathematically [23]:

$$\lambda(t) = \alpha\beta(\alpha t)^{\beta-1}e^{(\alpha t)^{\beta}}$$

(3.1)

where
   t is time.
   $\lambda(t)$ is the time-dependent failure rate (i.e., hazard rate).
   β is the shape parameter.
   α is the scale parameter.

At $\beta = 0.5$, Equation 3.1 gives the shape of the bathtub hazard rate curve shown in Figure 3.1.

3.10 General Reliability-Related Formulas

There are a number of general formulas used for performing various types of reliability-related analysis. Four of these formulas are presented below, separately.

3.10.1 Probability (or Failure) Density Function

The probability (or failure) density function is defined by

$$f(t) = -\frac{dR(t)}{dt}$$

(3.2)

where
   t is time.
   $f(t)$ is the probability (or failure) density function.
   $R(t)$ is the item/system reliability at time t.

EXAMPLE 3.1

Assume that the reliability of an oil and gas industry system is expressed by

$$R_{os}(t) = e^{-\lambda_{os}t}$$

(3.3)

where

$R_{os}(t)$ is the oil and gas industry system reliability at time t.
λ_{os} is the oil and gas industry system constant failure rate.

Obtain an expression for the probability (or failure) density function of the oil and gas industry system using Equation 3.2.
By substituting Equation 3.3 into Equation 3.2, we obtain

$$f(t) = -\frac{de^{-\lambda_{os}t}}{dt}$$

$$= \lambda_{os}e^{-\lambda_{os}t}$$

(3.4)

Thus, Equation 3.4 is the expression for the probability (or failure) density function of the oil and gas industry system.

3.10.2 Time-Dependent Failure Rate (or Hazard Rate) Function

The time-dependent failure rate (or hazard rate) function is defined by

$$\lambda(t) = \frac{f(t)}{R(t)}$$

(3.5)

where

$\lambda(t)$ is the system/item time-dependent failure rate (or hazard rate).

By substituting Equation 3.2 into Equation 3.5, we obtain

$$\lambda(t) = -\frac{1}{R(t)} \cdot \frac{dR(t)}{dt}$$

(3.6)

EXAMPLE 3.2

Obtain an expression for the time-dependent failure rate (or hazard rate) of the oil and gas industry system by using Equations 3.3 and 3.6.
By inserting Equation 3.3 into Equation 3.6, we obtain

$$\lambda(t) = -\frac{1}{e^{-\lambda_{os}t}} \cdot \frac{de^{-\lambda_{os}t}}{dt}$$

$$= \lambda_{os} \tag{3.7}$$

Thus, the time-dependent failure rate (or hazard rate) of the oil and gas industry system is given by Equation 3.7. It is to be noted that the right-hand side of this equation is not a function of time t. Needless to say, λ_{os} is usually referred to as the constant failure rate of an item/system (in this case of the oil and gas industry system) because it does not depend on time t.

3.10.3 General Reliability Function

This function can be obtained by using Equation 3.6. Thus, we get

$$-\lambda(t)dt = \frac{1}{R(t)} \cdot dR(t) \tag{3.8}$$

By integrating both sides of Equation 3.8 over the time interval $[0, t]$, we obtain

$$-\int_0^t \lambda(t)dt = \int_1^{R(t)} \frac{1}{R(t)} dR(t) \tag{3.9}$$

Since, at $t = 0$, $R(t) = 1$.

Evaluating the right-hand side of Equation 3.9 and then rearranging it yields

$$\ln R(t) = -\int_0^t \lambda(t)dt \tag{3.10}$$

Thus, from Equation 3.10, we get

$$R(t) = e^{-\int_0^t \lambda(t)dt} \tag{3.11}$$

Thus, Equation 3.11 is the general reliability function. This equation can be used to obtain the reliability function of an item/system when its times to failure follow any time-continuous probability distribution (e.g., exponential, Rayleigh, Weibull).

EXAMPLE 3.3

Assume that the hazard rate of an oil and gas industry system is expressed by Equation 3.1. Obtain an expression for the oil and gas industry system reliability function by using Equation 3.11.

By substituting Equation 3.1 into Equation 3.11, we obtain

$$R(t) = e^{-\int_0^t \left\{ \alpha\beta(\alpha t)^{\beta-1} e^{(\alpha t)^\beta} \right\} dt}$$

$$= e^{-\left\{ e^{(\alpha t)^\beta} - 1 \right\}} \tag{3.12}$$

Thus, Equation 3.12 is the expression for the oil and gas industry system reliability function.

3.10.4 Mean Time to Failure

Mean time to failure of an item/system can be obtained by using any of the following three formulas [6,24]:

$$MTTF = \int_0^\infty R(t)dt \tag{3.13}$$

or

$$MTTF = \lim_{s \to 0} R(s) \tag{3.14}$$

or

$$MTTF = E(t) = \int_0^\infty tf(t)dt \tag{3.15}$$

where
$MTTF$ is the mean time to failure of an item/system.
$E(t)$ is the expected value.
s is the Laplace transform variable.
$R(s)$ is the Laplace transform of the reliability function $R(t)$.

EXAMPLE 3.4

Using Equation 3.3, prove that Equations 3.13 and 3.14 yield the same result for the mean time to failure of the oil and gas industry system.

By inserting Equation 3.3 into Equation 3.13, we get

$$MTTF_{os} = \int_0^\infty e^{-\lambda_{os}t}\,dt$$

$$= \frac{1}{\lambda_{os}} \tag{3.16}$$

where
$MTTF_{os}$ is the mean time to failure of the oil and gas industry system.

By taking the Laplace transform of Equation 3.3, we obtain

$$R_{os}(s) = \int_0^\infty e^{-st} e^{-\lambda_{os}t}\,dt$$

$$= \frac{1}{1 + \lambda_{os}} \tag{3.17}$$

where
$R_{os}(s)$ is the Laplace transform of the oil and gas industry system reliability function $R_{os}(t)$.

By inserting Equation 3.17 into Equation 3.14, we obtain

$$MTTF_{os} = \lim_{s \to 0} \frac{1}{s + \lambda_{os}}$$

$$= \frac{1}{\lambda_{os}} \tag{3.18}$$

As Equations 3.16 and 3.18 are identical, it proves that Equations 3.13 and 3.14 yield the same result for the mean time to failure of the oil and gas industry systems.

3.11 Reliability Networks

A system in the area of oil and gas industry can form various types of configurations or networks in performing reliability analysis. Thus, this section is concerned with the reliability evaluation of such commonly occurring configurations or networks.

FIGURE 3.6
A *k*-unit series system (network).

3.11.1 Series Network

This is the simplest reliability network or configuration, and its block diagram is shown in Figure 3.6.

The diagram denotes a *k*-unit system, with each block in the diagram representing a unit. For the successful operation of the system, all *k* units must operate normally. In other words, if any one of the *k* units fails, the system fails.

In the series system, shown in Figure 3.6, reliability is expressed by

$$R_s = P(E_1 E_2 E_3 \dots E_k) \tag{3.19}$$

where

E_j is the successful operation (i.e., success event) of unit *j*; for *j* = 1, 2, 3, …, *k*.
$P(E_1 E_2 E_3 \dots E_k)$ is the occurrence probability of events E_1, E_2, E_3, …, E_k.
R_s is the series system reliability.

For independently failing units, Equation 3.19 becomes

$$R_s = P(E_1)P(E_2)P(E_3)\dots P(E_k) \tag{3.20}$$

where

$P(E_j)$ is the occurrence probability of event E_j, for *j* = 1, 2, 3, …, *k*.

If we let $R_j = P(E_j)$, for *j* = 1, 2, 3, …, *k*, Equation 3.20 becomes

$$R_s = R_1 R_2 R_3 \dots R_k$$

$$= \prod_{j=1}^{k} R_j \tag{3.21}$$

where

R_j is the unit *j* reliability; for *j* = 1, 2, 3, …, *k*.

For constant failure rate λ_j of unit *j* from Equation 3.11 (i.e., $\lambda_j(t) = \lambda_j$), we get

$$R_j(t) = e^{-\lambda_j t} \tag{3.22}$$

where
$R_j(t)$ is the reliability of unit j at time t.

By substituting Equation 3.22 into Equation 3.21, we obtain

$$R_s(t) = e^{-\sum\limits_{j=1}^{k} \lambda_j t} \tag{3.23}$$

where
$R_s(t)$ is the series system reliability at time t.

Substituting Equation 3.23 into Equation 3.13 yields the following expression for the series system mean time to failure:

$$MTTF_s = \int\limits_{0}^{\infty} e^{-\sum\limits_{j=1}^{k} \lambda_j t}\, dt$$

$$= \frac{1}{\sum\limits_{j=1}^{k} \lambda_j} \tag{3.24}$$

where
$MTTF_s$ is the series system mean time to failure.

By inserting Equation 3.23 into Equation 3.6, we obtain the following expression for the series system hazard rate:

$$\lambda_s(t) = -\frac{1}{e^{-\sum\limits_{j=1}^{k} \lambda_j t}} \left[-\sum\limits_{j=1}^{k} \lambda_j \right] e^{-\sum\limits_{j=1}^{k} \lambda_j t}$$

$$= \sum\limits_{j=1}^{k} \lambda_j \tag{3.25}$$

where
$\lambda_s(t)$ is the series system hazard rate.

It is to be noted that the right-hand side of Equation 3.25 is independent of time t. Thus, the left-hand side of this equation is simply λ_s, the series system failure rate. It means that whenever we add up failure rates of units/items, we automatically assume that these units/items fail independently and form a series network or configuration, a worst-case design scenario in regard to reliability.

EXAMPLE 3.5

Assume that an oil and gas industry system is composed of four identical and independently failing units, and the constant failure rate of each unit is 0.002 failures/h. All four units must operate normally for the oil and gas industry system to operate successfully. Calculate the following:

- The oil and gas industry system reliability for a 20-h mission
- The oil and gas industry system mean time to failure
- The oil and gas industry system failure rate

By inserting the given data values into Equation 3.23, we obtain

$$R_s(20) = e^{-(0.0002)(4)(20)}$$

$$= 0.9841$$

Using the given data values in Equation 3.24 yields

$$MTTF_s = \frac{1}{4(0.0002)} = 1250\,h$$

By substituting the specified data values into Equation 3.25, we get

$$\lambda_s = 4(0.0002)$$

$$= 0.0008\ \text{failures/h}$$

Thus, the oil and gas industry system reliability, mean time to failure, and failure rate are 0.9841, 1250 h, and 0.0008 failures/h, respectively.

3.11.2 Parallel Network

In this case, the system/network has k simultaneously operating units/items, and at least one of these units/items must operate normally for the system's successful operation. The k-unit parallel system block diagram is shown in Figure 3.7, and each block in the diagram denotes a unit/item.

The failure probability of the parallel system shown in Figure 3.7 is expressed by

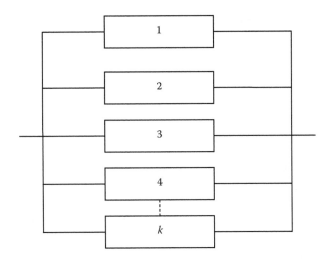

FIGURE 3.7
Block diagram of a parallel system with k units.

$$F_p = P\left(\bar{E}_1\bar{E}_2\bar{E}_3\bar{E}_4\ldots\bar{E}_k\right) \qquad (3.26)$$

where
$\bar{E}_j$ is the failure (i.e., failure event) of unit j, for $j = 1, 2, 3, 4, \ldots, k$.
$P(\bar{E}_1\bar{E}_2\bar{E}_3\bar{E}_4\ldots\bar{E}_k)$ is the occurrence probability of events $\bar{E}_1, \bar{E}_2, \bar{E}_3, \bar{E}_4, \ldots,$
and $\bar{E}_k$.
F_p is the failure probability of the parallel system.

For independently failing parallel units, Equation 3.26 becomes

$$F_p = P(\bar{E}_1)P(\bar{E}_2)P(\bar{E}_3)P(\bar{E}_4)\ldots P(\bar{E}_k) \qquad (3.27)$$

where
$P(\bar{E}_j)$ is the probability of occurrence of failure event $\bar{E}_j$, for $j = 1, 2, 3, 4, \ldots, k$.

If we let $F_j = P(\bar{E}_j)$ for $j = 1, 2, 3, 4, \ldots, k$, then Equation 3.27 becomes

$$F_p = F_1F_2F_3F_4\ldots F_k \qquad (3.28)$$

where
F_j is the unit j probability of failure for $j = 1, 2, 3, 4, \ldots, k$.

By subtracting Equation 3.28 from unity, we get

$$R_p = 1 - F_p$$

$$= 1 - F_1 F_2 F_3 F_4 \ldots F_k \tag{3.29}$$

where
R_p is the parallel system reliability.

For constant failure rate λ_j of unit j, subtracting Equation 3.22 from unity and then inserting it into Equation 3.29 yields

$$R_p(t) = 1 - \left(1 - e^{-\lambda_1 t}\right)\left(1 - e^{-\lambda_2 t}\right)\left(1 - e^{-\lambda_3 t}\right)\left(1 - e^{-\lambda_4 t}\right)\ldots\left(1 - e^{-\lambda_k t}\right) \tag{3.30}$$

where
$R_p(t)$ is the parallel system reliability at time t.

For identical units, by inserting Equation 3.30 into Equation 3.13, we get

$$MTTF_p = \int_0^\infty \left[1 - \left(1 - e^{-\lambda t}\right)^k\right] dt$$

$$= \frac{1}{\lambda} \sum_{j=1}^{k} \frac{1}{j} \tag{3.31}$$

where
λ is the unit constant failure rate.
$MTTF_p$ is the parallel system mean time to failure.

EXAMPLE 3.6

Assume that an oil and gas industry system is composed of two identical and independently failing units, and the constant failure rate of each unit is 0.0005 failures/h. At least one unit must operate normally for the oil and gas industry system to operate successfully.

Calculate the oil and gas industry system mean time to failure and reliability for a 200-h mission.

By substituting the given data values into Equation 3.30, we get

$$R_p(200) = 1 - \left[1 - e^{-(.0005)(200)}\right]^2$$

$$= 0.9909$$

Using the specified data values in Equation 3.31 yields

$$MTTF_p = \frac{1}{0.0005}\left[1+\frac{1}{2}\right]$$

$$= 3000\,h$$

Thus, the oil and gas industry system mean time to failure and reliability for a 200-h mission are 3000 h and 0.9909, respectively.

3.11.3 *n*-out-of-*k* Network

This is another type of redundancy in which at least *n* units out of a total of *k* active units must operate normally for the successful system/network operation. The block diagram of an *n*-out-of-*k* unit system/network is shown in Figure 3.8. The parallel and series networks are special cases of this network for $n = 1$ and $n = k$, respectively.

With the aid of binomial distribution, for identical and independent units, we write down the following expression for reliability of *n*-out-of-*k* unit network shown in Figure 3.8:

$$R_{n/k} = \sum_{j=n}^{k}\binom{k}{j}R^{j}\left(1-R\right)^{k-j} \tag{3.32}$$

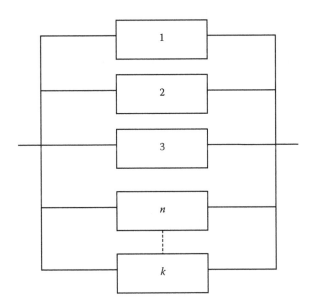

FIGURE 3.8
n-out-of-*k* unit network diagram.

where

$$\binom{k}{j} = \frac{k!}{(k-j)!j!}$$ (3.33)

R is the unit reliability.

$R_{n/k}$ is the n-out-of-k network reliability.

For identical units and their constant failure rates, using Equations 3.11 and 3.32, we get

$$R_{n/k}(t) = \sum_{j=n}^{k}\binom{k}{j}e^{-j\lambda t}\left(1-e^{-\lambda t}\right)^{k-j}$$ (3.34)

where

$R_{n/k}(t)$ is the n-out-of-k network reliability a time t.

λ is the unit constant failure rate.

By inserting Equation 3.34 into Equation 3.13, we get

$$MTTF_{n/k} = \int_0^{\infty}\left[\sum_{j=n}^{k}\binom{k}{j}e^{-j\lambda t}\left(1-e^{-\lambda t}\right)^{k-j}\right]dt$$

$$= \frac{1}{\lambda}\sum_{j=n}^{k}\frac{1}{j}$$ (3.35)

where

$MTTF_{n/k}$ is the n-out-of-k network mean time to failure.

EXAMPLE 3.7

Assume that an oil and gas industry system has three identical and independent units operating in parallel. At least two units must operate normally for the successful operation of the oil and gas industry system. Calculate the oil and gas industry system mean time to failure if the unit constant failure rate is 0.0004 failures/h.

By inserting the given data values into Equation 3.35, we obtain

$$MTTF_{2/3} = \frac{1}{(0.0004)}\sum_{j=2}^{3}\frac{1}{j}$$

$$= \frac{1}{(0.0004)}\left[\frac{1}{2}+\frac{1}{3}\right]$$

$$= 2083.33 \text{ h}$$

Thus, the oil and gas industry system mean time to failure is 2083.33 h.

3.11.4 Standby System

This is another type of reliability network/configuration in which only one unit operates and k units are kept in their standby mode. The network/system contains a total of $(k+1)$ units; as soon as the operating unit fails, the switching mechanism detects the failure and turns on one of the standby units. The system fails when all the standby units fail. The block diagram of a standby system with one operating and k standby units is shown in Figure 3.9.

With the aid of Figure 3.9 diagram, for independent and identical units, time-dependent unit failure rate, and perfect switching mechanism and standby units, we write down the following equation for the standby system reliability [25]:

$$R_{ss}(t) = \sum_{j=0}^{k} \left[\frac{\left[\int_0^t \lambda(t)\,dt \right]^j e^{-\int_0^t \lambda(t)\,dt}}{j!} \right] \tag{3.36}$$

where
  $R_{ss}(t)$ is the standby system reliability at time t.
  $\lambda(t)$ is the unit hazard rate or time-dependent failure rate.

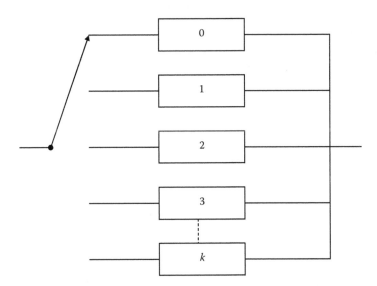

FIGURE 3.9
Block diagram of a standby system with one operating and k standby units.

For constant unit failure rate (i.e., $\lambda = \lambda(t)$), Equation 3.36 becomes

$$R_{ss}(t) = \sum_{j=0}^{k}\left[\frac{(\lambda t)^j e^{-\lambda t}}{j!}\right] \qquad (3.37)$$

where
   λ is the unit constant failure rate.

By inserting Equation 3.37 into Equation 3.13, we get

$$MTTF_{ss} = \int_{0}^{\infty}\left[\sum_{j=0}^{k}\left[\frac{(\lambda t)^j e^{-\lambda t}}{j!}\right]\right]dt$$

$$= \frac{k+1}{\lambda} \qquad (3.38)$$

where
   $MTTF_{ss}$ is the standby system mean time to failure.

EXAMPLE 3.8

A standby system is composed of two independent and identical units (one operating, the other on standby). The unit constant failure rate is 0.008 failures/h.

Calculate the standby system reliability for a 200-h mission by assuming that the switching mechanism is perfect and the standby unit remains as good as new in its standby mode.

By substituting the specified data values into Equation 3.37, we get

$$R_{ss}(200) = \sum_{j=0}^{1}\frac{\left[\left[(0.008)(200)\right]^j e^{-(0.008)(200)}\right]}{j!}$$

$$= 0.5249$$

Thus, the standby system reliability for the specified time period is 0.5249.

PROBLEMS

   1. Describe at least seven safety management principles.
   2. Discuss product hazard classifications.

3. What are the common causes of work-related injuries?

4. Describe the human factors accident causation theory.

5. What are the classifications of occupational stressors that can compromise safety?

6. List at least 12 reasons for the occurrence of human errors.

7. Describe the bathtub hazard rate curve.

8. What are the special case networks of the *n*-out-of-*k* network?

9. Write down general equations for the following:
   - Reliability function
   - Hazard rate
   - Failure density function

10. Write down three different formulas that can be used to develop expressions for mean time to failure.

References

1. Goetsch, D.L., *Occupational Safety and Health*, Prentice-Hall, Englewood Cliffs, New Jersey, 1996.
2. Dhillon, B.S., *Engineering Safety: Fundamentals, Techniques, and Applications*, World Scientific Publishing, River Edge, New Jersey, 2003.
3. Hammer, W., Price, D., *Occupational Safety Management and Engineering*, Prentice-Hall, Upper Saddle River, New Jersey, 2001.
4. Lyman, W.J., Fundamental consideration in preparing a master plan, *Electrical World*, 101, 1933, 778–792.
5. Smith, S.A., Service reliability measured by probabilities of outage, *Electrical World*, 103, 1934, 371–374.
6. Dhillon, B.S., *Design Reliability: Fundamentals and Applications*, CRC Press, Boca Raton, Florida, 1999.
7. Ladon, J., Editor, *Introduction to Occupational Health and Safety*, National Safety Council, Chicago, Illinois, 1986.
8. Dhillon, B.S., *Reliability, Quality, and Safety for Engineers*, CRC Press, Boca Raton, Florida, 2005.
9. Accidental Facts, Report, National Safety Council, Chicago, Illinois, 1996.
10. Petersen, D., *Safety Management, American Society of Safety Engineers*, Des Plaines, Illinois, 1998.
11. Hunter, T.A., *Engineering Design for Safety*, McGraw-Hill Book Company, New York, 1992.
12. Hammer, W., *Product Safety Management and Engineering*, Prentice-Hall, Englewood Cliffs, New Jersey, 1980.
13. Dhillon, B.S., *Robot System Reliability and Safety: A Modern Approach*, CRC Press, Boca Raton, Florida, 2015.

14. Dhillon, B.S., *Transportation Systems Reliability and Safety*, CRC Press, Boca Raton, Florida, 2011.
15. Heinrich, H.W., Petersen, D., Roos, N., *Industrial Accident Prevention*, McGraw-Hill Book Company, New York, 1980.
16. Beech, H.R., Burns, L.E., Sheffield, B.F., *A Behavioural Approach to the Management of Stress*, John Wiley and Sons, New York, 1982.
17. Meister, D., The problem of human-initiated failures, *Proceedings of the 8th National Symposium on Reliability and Quality Control*, 1962, pp. 234–239.
18. Juran, J.M., Inspector's errors in quality control, *Mechanical Engineering*, 57, 1935, 643–644.
19. McCormack, R.L., Inspection Accuracy: A Study of the Literature, Report No. SCTM 53-61(14), Sandia Corporation, Albuquerque, New Mexico, 1961.
20. Meister, D., *Human Factors: Theory and Practice*, John Wiley and Sons, New York, 1971.
21. Dhillon, B.S., *Human Reliability: With Human Factors*, Pergamon Press, New York, 1986.
22. Kapur, K.C., Reliability and Maintainability, in *Handbook of Industrial Engineering*, edited by G. Salvendy, John Wiley and Sons, New York, 1982, pp. 8.5.1–8.5.34.
23. Dhillon, B.S., Life distributions, *IEEE Transactions on Reliability*, 30(5), 1981, 457–460.
24. Shooman, M.L., *Probabilistic Reliability: An Engineering Approach*, McGraw-Hill Book Company, New York, 1968.
25. Sandler, G.H., *System Reliability Engineering*, Prentice-Hall, Englewood Cliffs, New Jersey, 1963.

4

Methods for Performing Safety and Reliability Analyses in the Oil and Gas Industry

4.1 Introduction

Over the years, a large amount of published literature on various aspects of safety and reliability has appeared in the form of books, technical reports, conference proceedings articles, and journal articles [1–4]. Many of these publications report the development of various types of methods and approaches for performing safety and reliability analyses. Some of these methods and approaches can be used for performing analysis in both safety and reliability fields. The others are more confined to a specific field (e.g., safety or reliability).

Fault tree analysis (FTA), failure modes and effect analysis (FMEA), and the Markov method are the examples of methods that can be used in both safety and reliability fields. The FTA method was developed in the early 1960s for analyzing the safety of rocket launch control systems, and FMEA was developed in the early 1950s for analyzing the reliability of engineering systems.

The Markov method is named after the Russian mathematician Andrei A. Markov (1856–1922) and is a highly mathematical approach that is frequently used for performing various types of safety and reliability analyses in engineering systems. This chapter presents a number of methods and approaches extracted from the published literature, considered useful to perform safety and reliability analyses in the oil and gas industry.

4.2 Root Cause Analysis

Root cause analysis (RCA) may simply be described as a systematic investigation approach that makes use of information collected during an assessment of an accident, for determining the underlying factors for deficiencies that caused the accident [5,6]. It was developed by the U.S. Department of

Energy for investigating incidents [7]. The following 10 general steps are involved in performing RCA [7–9]:

- *Step 1*: Educating all personnel involved in RCA
- *Step 2*: Informing all appropriate staff members when a sentinel event is reported
- *Step 3*: Forming an RCA team made up of appropriate personnel
- *Step 4*: Preparing for and holding the first team meeting
- *Step 5*: Determining the event sequence
- *Step 6*: Separating and identifying each event sequence that may have been a contributory factor in the occurrence of the sentinel event
- *Step 7*: Brainstorming about the factors surrounding the chosen events that may have been contributory to the occurrence of the sentinel event
- *Step 8*: Affinitizing with the results of the brainstorming session
- *Step 9*: Developing the action plan
- *Step 10*: Distributing the action plan and the RCA document to all concerned personnel

Some of the benefits and drawbacks of the RCA method are as follows [7,9,10]:

- *Benefits*
  - It is an effective tool to highlight and address systems and organization-related issues.
  - It is a well-structured and process-focused method.
  - The systematic application of this method can identify common root causes that link a disparate collection of accidents.
- *Drawbacks*
  - In essence, this method is basically an uncontrolled case study.
  - It is quite a time-consuming and labor-intensive approach.
  - The method is possible to be tainted by hindsight bias.
  - It is impossible to determine exactly if the root cause highlighted by the analysis is really the cause for the accident occurrence.

4.3 Hazards and Operability Analysis

Hazard and operability analysis (HAZOP) is a systematic method for identifying hazards and operating problems in a facility. Over the years, HAZOP has proven to be an extremely useful method for identifying unforeseen hazards designed into facilities due to various reasons, or introduced into existing facilities due to factors such as changes made to operating procedures or process conditions.

Three fundamental objectives of HAZOP are as follows [11,12]:

1. To produce a complete facility/process description
2. To review each and every process/facility part to discover how deviations from the design intentions can happen
3. To decide whether deviations from design intentions can result in operating hazards/problems

A HAZOP study can be conducted in five steps shown in Figure 4.1 [11]. Each of these steps is described below in more detail.

Step 1 is concerned with establishing study scope and objectives by considering all involved relevant factors. Step 2 is concerned with forming the HAZOP team by ensuring that the team comprises personnel from the area of design and operation with appropriate experiences for determining the effects of deviations from the anticipated application.

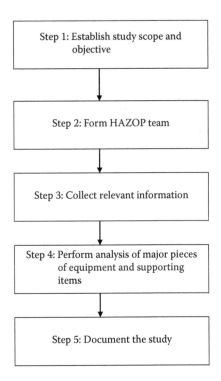

FIGURE 4.1
HAZOP study steps.

Step 3 is concerned with obtaining the necessary documentation, process description, and drawings, including items such as equipment specifications, process flow sheets, process-control logic diagrams, layout drawings, emergency response procedures, and operating and maintenance procedures. Step 4 is concerned with performing analysis of all major pieces of equipment and of all supporting equipment, instrumentation, and piping with the aid of step 3 documents.

Finally, step 5 is concerned with documenting the consequences of any deviation from the norm, and summary of deviations from the norm and a summary of those deviations considered credible and quite hazardous.

4.4 Technique of Operations Review

Technique of operations review (TOR) was developed by D.A. Weaver of the American Society of Safety Engineers in the early 1970s [13]. In regard to safety, it seeks to highlight systemic causes for the occurrence of an adverse incident rather than assigning blame. Furthermore, the method allows management personnel and workers to work jointly to analyze workplace-related accidents, incidents, and failures. Thus, TOR may simply be described as a hands-on analytical method to highlight the root system causes of an operation failure [13,14].

The method makes use of a worksheet containing simple terms that require yes/no decisions and is activated by an adverse incident occurring at a certain location and time involving certain people. It is to be noted that this method is not a hypothetical process and demands a systemic evaluation of the circumstances surrounding the incident in question [12]. Ultimately, TOR highlights how the company/organization could have prevented the occurrence of the accident.

TOR is composed of the following eight steps [12,14,15]:

- *Step 1: Form the TOR team.* This step is concerned with forming a TOR team with members from all concerned areas.
- *Step 2: Hold a roundtable session.* This step is concerned with holding a roundtable session to impart common knowledge to all members of the TOR team.
- *Step 3: Identify one key systemic factor that was instrumental in causing the accident/incident.* This step is concerned with identifying one key systemic factor that played an instrumental role in causing the accident/incident. It is to be noted that this factor must be based on all team members' consensus and serves as a starting point for further investigation.

- *Step 4: Use the team consensus.* This step is concerned with using the team consensus when responding to a sequence of yes/no options.
- *Step 5: Evaluate the identified factors.* This step is concerned with evaluating the identified factors and ensuring that there is consensus among the members of the team with respect to each factor.
- *Step 6: Prioritize the contributory factors.* This step is concerned with prioritizing all the contributory factors by starting with the most serious one.
- *Step 7: Develop necessary corrective/preventive strategies.* This step is concerned with developing necessary corrective/preventive strategies with respect to each contributory factor.
- *Step 8: Execute strategies' implementation.* This step is concerned with carrying out the implementation of the strategies.

Finally, it is added that the main strength of this method (i.e., TOR) is the involvement of the line personnel in the analysis, and its main weakness is that it is an after-the-fact process.

4.5 Interface Safety Analysis

Interface safety analysis (ISA) is concerned with determining the incompatibilities between subsystems of an equipment/product and assemblies that could cause accidents. The method establishes that it is possible to integrate distinct units/parts into a viable system and that normal operation of an individual part or unit will not impair the performance of or damage another unit/part or the total system/equipment.

The relationships considered by ISA can be grouped under three classifications: physical, flow, and functional [12,14,16]. Each of these classifications of relationships is described below, separately.

4.5.1 Physical Relationships

These relationships are concerned with the products'/items' physical aspects. For example, two products/items could be well designed and manufactured and operate well individually, but they may have difficulties in fitting together due to dimension-related differences, or there could be other incompatibilities that may result in safety-related problems. Examples of the other problems are as follows:

- A very small clearance between units, thus during the removal process, the units could be damaged

- Restricted or impossible access to or egress from equipment/system
- Impossible to mate, tighten, or join parts properly

4.5.2 Flow Relationships

These relationships are concerned with the flow between two or more units/items. For example, the flow between two units/items may involve steam, water, electrical energy, fuel, air, or lubricating oil. Furthermore, the flow could also be unconfined, such as heat radiation from one body/unit/item to another. The problems experienced with many products quite often include the proper flow of fluids and energy from one item/unit to another item/unit through confined passages, consequently leading to safety-related problems.

The causes of flow-related problems include complete or partial interconnection failure and faulty connections between units/items. In regard to fluids, the factors that must be considered with care from the safety aspect include loss of pressure, contamination, flammability, lubricity, odor, and toxicity.

4.5.3 Functional Relationships

These are concerned with function-related relationships with multiple items/units. For example, in a situation where outputs of a unit/item constitute the inputs to the downstream item(s)/unit(s), any error in outputs and inputs may result in damage to the downstream item(s)/unit(s), thereby creating a safety hazard or problem. Such outputs could be in conditions such as excessive, degraded, erratic, zero, and unprogrammed outputs.

4.6 Job Safety Analysis

Job safety analysis (JSA) is concerned with uncovering and rectifying potential hazards intrinsic to or inherent in the workplace. Usually, workers, supervisors, safety professionals, and management participate in JSA. JSA is composed of the following five steps [1,12]:

- *Step 1*: Selecting a job for analysis
- *Step 2*: Breaking down the job into a number of steps or tasks
- *Step 3*: Identifying all potential hazards and determining the appropriate actions to control these hazards
- *Step 4*: Applying the actions to control the hazards
- *Step 5*: Evaluating the controls

Finally, it is added that the success of this method very much depends on the degree of the rigor the JSA team members exercise during analysis. Additional information on this method is available in Reference 1.

4.7 Preliminary Hazard Analysis

Preliminary hazard analysis (PHA) is widely used during the concept design phase. It is an unstructured approach and is used in situations when there is a lack of definitive information, such as drawings and functional flow diagrams. Over the years, the method has proved to be a very good tool for taking early appropriate measures to highlight and eliminate possible hazards when all the required data are not available. The findings of the method are considered to be quite useful for serving as a guide in potential detailed analysis.

The method requires the formation of an ad hoc team of personnel with appropriate level of familiarity with items such as equipment, substances, materials, and/or the process under consideration. The members of the team review with care the hazards' occurrence in the area of their expertise and experience, and as a team they play the devil's advocate.

Additional information on PHA is available in Reference 1.

4.8 Failure Modes and Effect Analysis

Failure modes and effect analysis (FMEA) is a widely used design tool for analyzing the reliability of engineering systems. It may simply be described as an approach to analyze the effects of potential failure modes in the system [3]. The history of FMEA goes back to the early years of the 1950s with the development of flight control systems, when the Bureau of Aeronautics of the U.S. Navy developed a requirement called "failure analysis" for establishing a mechanism for reliability control over the detail design-related effort [17]. Subsequently, the term "failure analysis" was switched over to "failure modes and effect analysis" (FMEA).

The seven main steps that are usually followed to perform FMEA are as follows [3,12]:

- *Step 1*: Define system boundaries and detailed requirements
- *Step 2*: List system subsystems and parts/components
- *Step 3*: List each part's/component's failure modes, the identification, and the description

- *Step 4*: Assign failure occurrence probability/rate to each part/ component failure mode
- *Step 5*: List effect or effects of each failure mode on subsystem(s), system, plant
- *Step 6*: Enter remarks for each and every failure mode
- *Step 7*: Review each and every critical failure mode and take necessary actions

There are many factors that must be explored carefully before the implementation of FMEA. Some of these factors are as follows [12,18,19]:

- Measuring advantages/costs
- Examination of each and every conceivable failure mode by the involved professionals only
- Making decisions on the basis of the risk priority number
- Obtaining approval and support of the engineer

Over the years, professionals working in the area of reliability engineering have established certain facts/guidelines concerning this method (i.e., FMEA). Four of these facts/guidelines are as follows [12,19]:

1. FMEA is not a method to select the optimum design concept.
2. FMEA has certain limitations.
3. FMEA is not designed for superseding the work of the engineer.
4. Do not develop the majority of FMEA in a meeting.

Some of the main benefits of performing FMEA are as follows [3,12,18,19]:

- A useful approach for comparing designs and identifying safety-related concerns
- A systematic approach for classifying hardware failures
- A useful approach for understanding and improving customer satisfaction
- A useful method that starts from the detailed level and works upward
- A useful approach for improving communication between design interface personnel
- A useful method for safeguarding against repeating the same mistakes in the future
- A visibility approach for management that cuts down product development cost and time
- A useful method for reducing engineering-related changes and for improving the efficiency of test planning

4.8.1 Failure Mode Effects and Criticality Analysis

Failure mode effects and criticality analysis (FMECA) is an extended version of FMEA. More clearly, when FMEA is extended to group or categorize each failure effect with respect to its level of severity (this includes documenting catastrophic and critical failures), then it (i.e., FMEA) is called FMECA. It was developed by the National Aeronautics and Astronautics Administration (NASA) for assuring the required reliability of space systems. A military standard titled "Procedures for Performing a Failure Mode, Effects, and Criticality Analysis" was developed by the U.S. Department of Defense in the 1970s [20].

In order to perform FMECA effectively, various types of information are required. In particular, the design-related information required for the FMECA includes the following items [3,12]:

- Design descriptions
- System schematics
- Equipment/part drawings
- Functional block diagrams
- Operating specifications and limitations
- Field service data
- Interface specifications
- Relevant specifications (i.e., company, customer, etc.)
- Configuration management-related data
- Reliability data
- Effects of environment on item under consideration

Additional information on FMEA/FMECA is available in Reference 3.

4.9 Fault Tree Analysis

Fault tree analysis (FTA) is a widely used method in industry for evaluating the reliability of engineering systems during their design and development phase, particularly in the area of nuclear power generation. A fault tree may simply be described as a logical representation of the basic fault events that lead to a specified undesirable event, called the "top event," and is depicted using a upside down tree structure with logic gates such as OR and AND.

The method was developed in the early 1960s at the Bell Telephone Laboratories for performing analysis of the Minuteman Launch Control

System [2,3]. Some of the main objectives of conducting FTA are as follows [3]:

- To confirm the system's ability to meet its imposed safety requirements
- To understand the system failures' functional relationship
- To comprehend the degree of protection that the design concept provides against the occurrence of failures
- To highlight critical areas and cost-effective improvements
- To meet jurisdictional-related requirements

There are many prerequisites associated with this method. Some of these prerequisites are as follows [3,12]:

- Clearly defined objectives and scope of analysis
- Clearly defined system interfaces and physical bounds
- Clear understanding of design, operation, and maintenance aspects of the system under consideration
- A thorough review of system operation-related experience
- Clear definition of what constitutes system failure (i.e., undesirable event)
- Clear identification of all associated assumptions

FTA begins by identifying an undesirable event, called top event, associated with a system under consideration. Fault events that can cause the top event occurrence are generated and connected by logic operators such as OR and AND. The operator/gate OR provides a true output (i.e., fault) when one or more inputs (i.e., input faults) are true. Similarly, the operator/gate AND provides a true output (i.e., fault) when all the inputs (i.e., input faults) are true.

A fault tree's construction proceeds by generating fault events in a successive manner until the fault events need not be developed any further. These fault events are known as primary or basic fault events. During the construction of a fault tree, a question that is asked over and over again is, "How could this fault event occur?"

The basic symbols used to construct fault trees are shown in Figure 4.2. All of these symbols are described as follows:

- *Circle*: It denotes a basic or primary fault event (e.g., failure of an elementary part). The fault event's occurrence probability, failure, and repair rates are normally obtained from empirical data.
- *Rectangle*: It represents a fault event that results from the logical combination of fault events through the input of a logic gate such as AND and OR.

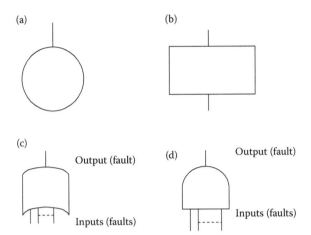

FIGURE 4.2
Basic fault tree symbols: (a) basic fault event, (b) resultant event, (c) OR gate, (d) AND gate.

- *OR gate*: It denotes that an output fault event occurs when one or more of the input fault events occur.
- *AND gate*: It denotes that an output fault event occurs if and only if all of the input fault events occur.

EXAMPLE 4.1

A windowless room has a switch and two light bulbs. Develop a fault tree for the undesired or top fault event "dark room" (i.e., room without light). Assume that the room can only be dark if the switch fails to close, both the light bulbs have burnt out, or there is no electricity. A fault tree for this example is shown in Figure 4.3. The single capital letters in the diagram denote corresponding fault events (e.g., *A*: fuse failure, *F*: no electricity, and *T*: dark room).

4.9.1 Probability Evaluation of Fault Trees

When the probabilities of occurrence of primary/basic fault events are known, the occurrence probability of the top event can be calculated. This can only be achieved by first calculating probabilities of occurrence of the output fault events of all the intermediate and lower logic gates (e.g., AND and OR gates).

Thus, the probability of occurrence of the OR gate output fault, *Y*, is given by [2,3]

$$P(Y) = 1 - \prod_{j=1}^{n}\left\{1 - P(Y_j)\right\} \tag{4.1}$$

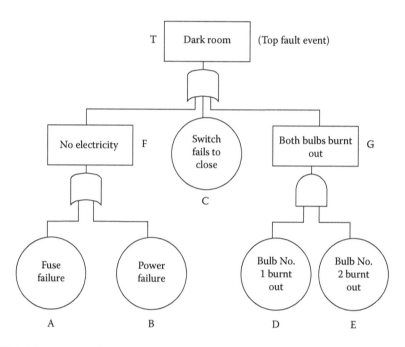

FIGURE 4.3
A fault tree for the top fault event: dark room.

where
  $P(Y)$ is the probability of occurrence of the OR gate output fault event Y.
  n is the total number of OR gate input fault events.
  $P(Y_j)$ is the probability of occurrence of the OR gate input fault event Y_j for $j = 1, 2, 3, ..., n$.

Similarly, the probability of occurrence of the AND gate output fault event, X, is expressed by [3,21]

$$P(X) = \prod_{j=1}^{m} P(X_j) \tag{4.2}$$

where
  $P(X)$ is the probability of occurrence of the AND gate output fault event X.
  m is the total number of AND gate input fault events.
  $P(X_j)$ is the occurrence probability of the AND gate input fault event X_j, for $j = 1, 2, 3, ..., m$.

EXAMPLE 4.2

Assume that the probabilities of occurrence of fault events, switch fails to close, bulb No. 1 burnt out, bulb No. 2 burnt out, fuse failure, and power failure in Figure 4.3 are 0.06, 0.05, 0.04, 0.03, and 0.02, respectively. Calculate the probability of occurrence of the top fault event "dark room" with the aid of Equations 4.1 and 4.2.

Using the given data values of the fault events A and B and Equation 4.1, we get

$$P(F) = 1 - (1 - P(A))(1 - P(B))$$

$$= 1 - (1 - 0.03)(1 - 0.02)$$

$$= 0.0494$$

where

$P(F)$ is the probability of occurrence of fault event F.
$P(A)$ is the probability of occurrence of fault event A.
$P(B)$ is the probability of occurrence of fault event B.

Similarly, using the given data values of the fault events D and E and Equation 4.2, we obtain

$$P(G) = P(D)\, P(E)$$

$$= (0.05)(0.04)$$

$$= 0.002$$

where

$P(G)$ is the probability of occurrence of fault event G.
$P(D)$ is the probability of occurrence of fault event D.
$P(E)$ is the probability of occurrence of fault event E.

By inserting the above two calculated values and the given data value into Equation 4.1, we obtain

$$P(T) = 1 - (1 - P(F))(1 - P(C))(1 - P(G))$$

$$= 1 - (1 - 0.0494)(1 - 0.06)(1 - 0.002)$$

$$= 0.1082$$

where

$P(T)$ is the probability of occurrence of the top fault event T.
$P(C)$ is the probability of occurrence of fault event C.

Thus, the probability of occurrence of the top fault event T (dark room) is 0.1082.

4.9.2 Fault Tree Analysis: Advantages and Disadvantages

There are many advantages and disadvantages of the FTA. Some of its advantages are as follows [2,3]:

- Useful to provide insight into the system behavior
- Useful to handle complex systems more easily
- A graphic aid for management
- Useful to identify failures deductively
- Useful because it allows concentration on one particular failure at a time
- Useful to provide options for management and others to perform either qualitative or quantitative analysis
- Useful because it requires the analyst to understand thoroughly the system under consideration before starting the analysis

In contrast, some of the disadvantages of the FTA are as follows [2,3]:

- It is a time-consuming and costly method.
- End results are quite difficult to check.
- It considers parts/components in either a working state or a failed state. More specifically, the partial failure states of the parts/components are difficult to handle.

Additional information on FTA is available in References 2 and 3.

4.10 Markov Method

This method is often used to perform reliability-related analysis of repairable systems with constant failure and repair rates. The following assumptions are associated with the Markov method [3,21]:

- All occurrences are independent of each other.
- The transitional probability from one system state to another in the finite time interval Δt is given by $\alpha \Delta t$, where α is the transition rate (e.g., failure or repair rate) from one system state to another.
- The probability of more than one transition occurrence in the finite time interval Δt from one system state to another is negligible (e.g., $(\alpha \Delta t)(\alpha \Delta t) \rightarrow 0$).

The following example demonstrates the application of the Markov method.

EXAMPLE 4.3

Assume that a system used in the oil and gas industry can either be in an operating state or a failed state. The constant failure rate and the

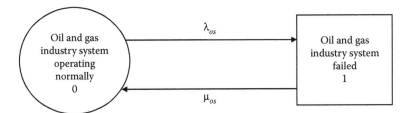

FIGURE 4.4
Oil and gas industry system state space diagram.

repair rate of the system are λ_{os} and μ_{os}, respectively. The system state space diagram is shown in Figure 4.4. The numerals in circle and box denote the system states. Develop expressions for the oil and gas industry system time-dependent and steady-state availabilities and unavailabilities, reliability, and mean time to failure with the aid of Markov method.

With the aid of Markov method, we write down the following equations for states 0 and 1 shown, respectively, in Figure 4.4.

$$P_0(t + \Delta t) = P_0(t)(1 - \lambda_{os}\Delta t) + P_1(t)\mu_{os}\Delta t \qquad (4.3)$$

$$P_1(t + \Delta t) = P_1(t)(1 - \mu_{os}\Delta t) + P_1(t)\lambda_{os}\Delta t \qquad (4.4)$$

where
t is time.
$\lambda_{os}\Delta t$ is the probability of the oil and gas industry system failure in finite time interval Δt.
$\mu_{os}\Delta t$ is the probability of the oil and gas industry system repair in finite time interval Δt.
$(1 - \lambda_{os}\Delta t)$ is the probability of no failure in finite time interval Δt.
$(1 - \mu_{os}\Delta t)$ is the probability of no repair in finite time interval Δt.
$P_j(t)$ is the probability that the oil and gas industry system is in state j at time t, for $j = 0, 1$.
$P_0(t + \Delta t)$ is the probability of the oil and gas industry system being in operating state 0 at time $(t + \Delta t)$.
$P_1(t + \Delta t)$ is the probability of the oil and gas industry system being in failed state 1 at time $(t + \Delta t)$.

From Equation 4.3, we obtain

$$P_0(t + \Delta t) = P_0(t) - P_0(t)\lambda_{os}\Delta t + P_1(t)\mu_{os}\Delta t \qquad (4.5)$$

From Equation 4.5, we write

$$\lim_{\Delta t \to 0} \frac{P_0(t + \Delta t) - P_0(t)}{\Delta t} = -P_0(t)\lambda_{os} + P_1(t)\mu_{os} \qquad (4.6)$$

Thus, from Equation 4.6, we get

$$\frac{dP_0(t)}{dt} + P_0(t)\lambda_{os} = P_1(t)\mu_{os} \tag{4.7}$$

Similarly, using Equation 4.4, we obtain

$$\frac{dP_1(t)}{dt} + P_1(t)\mu_{os} = P_0(t)\lambda_{os} \tag{4.8}$$

At time $t = 0$, $P_0(0) = 1$, and $P_1(0) = 0$.
Solving Equations 4.7 and 4.8, we get [3]

$$P_0(t) = \frac{\mu_{os}}{\left(\lambda_{os} + \mu_{os}\right)} + \frac{\lambda_{os}}{\left(\lambda_{os} + \mu_{os}\right)} e^{-(\lambda_{os} + \mu_{os})t} \tag{4.9}$$

$$P_1(t) = \frac{\lambda_{os}}{\left(\lambda_{os} + \mu_{os}\right)} - \frac{\lambda_{os}}{\left(\lambda_{os} + \mu_{os}\right)} e^{-(\lambda_{os} + \mu_{os})t} \tag{4.10}$$

Thus, the oil and gas industry system time-dependent availability and unavailability, respectively, are

$$A_{os}(t) = P_0(t) = \frac{\mu_{os}}{\left(\lambda_{os} + \mu_{os}\right)} + \frac{\lambda_{os}}{\left(\lambda_{os} + \mu_{os}\right)} e^{-(\lambda_{os} + \mu_{os})t} \tag{4.11}$$

and

$$UA_{os}(t) = P_1(t) = \frac{\lambda_{os}}{\left(\lambda_{os} + \mu_{os}\right)} - \frac{\lambda_{os}}{\left(\lambda_{os} + \mu_{os}\right)} e^{-(\lambda_{os} + \mu_{os})t} \tag{4.12}$$

where
$A_{os}(t)$ is the oil and gas industry system time-dependent availability.
$UA_{os}(t)$ is the oil and gas industry system time-dependent unavailability.

By letting time t go to ∞ in Equations 4.11 and 4.12, we obtain [3]

$$A_{os} = \lim_{t \to \infty} A_{os}(t) = \frac{\mu_{os}}{\lambda_{os} + \mu_{os}} \tag{4.13}$$

and

$$UA_{os} = \lim_{t \to \infty} UA_{os}(t) = \frac{\lambda_{os}}{\lambda_{os} + \mu_{os}} \tag{4.14}$$

where
A_{os} is the oil and gas industry system steady-state availability.
UA_{os} is the oil and gas industry system steady-state unavailability.

By setting $\mu_{os} = 0$ in Equation 4.9, we obtain

$$R_{os}(t) = P_0(t) = e^{-\lambda_{os}t} \qquad (4.15)$$

where
$R_{os}(t)$ is the oil and gas industry system reliability at time t.

By integrating Equation 4.15 over the time interval $[0, \infty]$, we obtain the following expression for the oil and gas industry system mean time to failure [3]:

$$MTTF_{os} = \int_0^\infty e^{-\lambda_{os}t} dt$$

$$= \frac{1}{\lambda_{os}} \qquad (4.16)$$

where
$MTTF_{os}$ is the oil and gas industry system mean time to failure.

Thus, the oil and gas industry system time-dependent and steady-state availabilities and unavailabilities, reliability, and mean time to failure are given by Equations 4.11, 4.12, 4.13, 4.14, 4.15, and 4.16, respectively.

EXAMPLE 4.4
Assume that the constant failure and repair rates of an oil and gas industry system are 0.0002 failures/h and 0.0004 repairs/h, respectively. Calculate the oil and gas industry system steady-state unavailability and unavailability during an 8-h mission.
By inserting the specified data values into Equations 4.14 and 4.12, we obtain

$$UA_{os} = \frac{0.0002}{0.0002 + 0.0004} = 0.3333$$

and

$$UA_{os}(8) = \frac{0.0002}{(0.0002 + 0.0004)} - \frac{0.0002}{(0.0002 + 0.0004)} e^{-(0.0002 + 0.0004)(8)}$$

$$= 0.0016$$

Thus, the oil and gas industry system steady-state unavailability and unavailability during an 8-h mission are 0.333 and 0.0016, respectively.

PROBLEMS

1. Describe the root cause analysis.
2. What are the benefits and drawbacks of fault tree analysis?
3. Describe the technique of operations review (TOR).
4. Prove Equations 4.9 and 4.10 by using Equations 4.7 and 4.8.
5. Compare hazards and operability analysis with interface safety analysis.
6. Describe failure modes and effect analysis.
7. What are the four basic symbols used to construct fault trees? Describe each of these symbols.
8. What is the difference between FMEA and FMECA?
9. What are the advantages and disadvantages of root cause analysis?
10. Assume that probabilities of occurrence of events A, B, C, D, and E in Figure 4.3 are 0.07, 0.05, 0.02, 0.08, and 0.04, respectively. Calculate the probability of occurrence of the top fault event T: dark room.

References

1. Hammer, W., Price, D., *Occupational Safety Management and Engineering*, Prentice-Hall, Upper Saddle River, New Jersey, 2001.
2. Dhillon, B.S., Singh, C., *Engineering Reliability: New Techniques and Applications*, John Wiley and Sons, New York, 1981.
3. Dhillon, B.S., *Design Reliability: Fundamentals and Applications*, CRC Press, Boca Raton, Florida, 1999.
4. Dhillon, B.S., *Human Reliability: With Human Factors*, Pergamon Press, New York, 1986.
5. Latino, R.J., Automating root cause analysis, in *Error Reduction in Health Care*, edited by P.L. Spath, John Wiley and Sons, New York, 2000, pp. 155–164.
6. Busse, D.K., Wright, D.J., Classification and Analysis of Incidents in Complex Medical Environments, Report, 2000. Available from the Intensive Care Unit, Western General Hospital, Edinburgh, UK.
7. Dhillon, B.S., *Safety and Human Error in Engineering Systems*, CRC Press, Boca Raton, Florida, 2013.
8. Burke, A., Root Cause Analysis, Report, 2000. Available from the Wild Iris Medical Education, PO Box 257, Comptche, California.
9. Dhillon, B.S., *Human Reliability and Error in Medical System*, World Scientific Publishing, River Edge, New Jersey, 2003.
10. Wald, H., Shojania, K.G., Root cause analysis, in *Making Health Care Safer: A Critical Analysis of Patient Safety Practices*, edited by A.J. Markowitz, Report No. 43, Agency for Health Care Research and Quality, U.S. Department of Health and Human Services, Rockville, Maryland, 2001, Chapter 5, pp. 1–7.

11. Risk Analysis Requirements and Guidelines, Document No. CAN/CSA-Q6340-91, Canadian Standards Association (CSA), 1991. Available from the Canadian Standards Association, Rexdale, Ontario, Canada.
12. Dhillon, B.S., *Transportation Systems Reliability and Safety*, CRC Press, Boca Raton, Florida, 2011.
13. Hallock, R.G., Technique of operations review analysis: Determines cause of accident/incident, *Safety and Health*, 60(8), 1991, 38–39.
14. Dhillon, B.S., *Engineering Safety: Fundamentals, Techniques, and Applications*, World Scientific Publishing, River Edge, New Jersey, 2003.
15. Goetsch, D.L., *Occupational Safety and Health*, Prentice-Hall, Englewood Cliffs, New Jersey, 1996.
16. Hammer, W., *Product Safety Management and Engineering*, Prentice-Hall, Englewood Cliffs, New Jersey, 1980.
17. Bureau of Naval Weapons, General Specification for Design, Installation, and Test of Aircraft Flight Control Systems, MIL-F-18372 (Aer), Bureau of Naval Weapons, Department of the Navy, Washington, DC.
18. McDermott, R.E., Mikulak, R.J., Beauregard, M.R., *The Basics of FMEA*, Quality Resources, New York, 1996.
19. Palady, P., *Failure Modes and Effects Analysis*, PT Publications, West Palm Beach, Florida, 1995.
20. DoD, Procedure for Performing a Failure Mode, Effects, and Criticality Analysis, MIL-STD-1629, Department of Defense, Washington, DC, 1980.
21. Shooman, M.L., *Probabilistic Reliability: An Engineering Approach*, McGraw-Hill Book Company, New York, 1968.

5

Safety in Offshore Oil and Gas Industry

5.1 Introduction

The history of the offshore oil and gas industry may be traced back to around 1891, when the first submerged oil wells were drilled from platforms on piles in the freshwaters of the Grand Lake St. Marys in Ohio, United States [1]. Over the past six decades, offshore production has increased tremendously, currently about 30% of world oil and gas production coming from offshore [2].

Offshore industry has become an important element of the industrial sector as each year a vast sum of money is spent on offshore-related developments around the globe. In order to meet the increasing demand for oil and gas, the industry uses and develops leading-edge technology to drill even deeper.

Over the years, many accidents in the offshore industrial sector have occurred and resulted in many fatalities and a large sum of money being spent on damages. Some examples of the deadliest accidents in the offshore oil and gas industry are the Piper Alpha platform accident in the United Kingdom in 1998, the Mumbai High North Platform accident in India in 2005, and the Alexander L. Kielland accident in Norway in 1980 [3].

Safety has become an important issue in the offshore oil and gas industry. This chapter presents various important aspects of safety in offshore oil and gas industry.

5.2 Offshore Industrial Sector Risk Picture

The offshore industrial sector "risk picture" is a quite multifaceted one. Therefore, to establish an actual "factual risk picture" is a difficult task due to factors such as follows [4]:

- Inconsistency in reporting and recording incidents across the industrial sector
- Different measures of risk

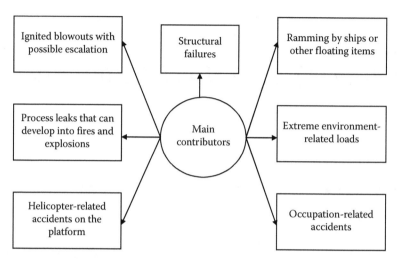

FIGURE 5.1
Main contributors to offshore individual risk.

- A basic uncertainty in extrapolating information from the past to present risk
- Changes in the regulatory regime and in the industrial sector and its management systems

Nonetheless, the main contributors to individual risk are shown in Figure 5.1 [4].

Additional information on offshore industrial sector risk picture is available in Reference 4.

5.3 Offshore Worker Situation Awareness Concept, Studies, and Their Results

Past experiences over the years clearly indicate that there is a definite need for maintaining the situation awareness of workers at a high level for insuring the safety of their operations in many industrial sectors. As per Reference 5, situation awareness may be expressed as the perception of the elements in the environment within the framework of a volume of time and space, the comprehension of their proper meanings, and the projection of their status in the near future. The following are the main elements of situation awareness [5]:

- *Situation awareness levels*: These are concerned with perception, comprehension, and projection. Perception calls for monitoring the surrounding environment on a continuous basis for encoding sensory-related information and detecting changes in significant stimuli.

Comprehension involves combination, interpretation, storage, and retention of incoming information to form a picture of the current situation or condition whereby the events'/objectives' significance is understood. Finally, projection is the result of perception and comprehension, and it is concerned with predicting possible future events/states.

- *Team situation awareness*: This is concerned with teamwork, as the successful accomplishment of a specified task (e.g., a drilling task in the offshore oil and gas industry) is completely dependent upon all the crew members collectively working together. Therefore, it is absolutely essential for all crew members to have a clear mutual understanding of the situation under consideration.

  More clearly and in short, it may simply be stated that all involved members of the crew must have a situation awareness (this shared awareness is known as team situation awareness) [3,5].

- *Factors affecting situation awareness*: Two factors that affect situation awareness are workload and stress. Unusually high or low workloads are considered to potentially impact performance of humans to a certain degree [6]. Low workload can result in boredom with consequent inattentiveness, significantly reduced motivation, and lower vigilance. Furthermore, when less attention is being given to workplace conditions or situations that, in turn, can lead to poor situation awareness.

  On the other hand, high workloads can lead to impairing situation awareness of workers as they may not be completely aware of situation-related changes, and can make wrong decisions on the basis of incorrect or incomplete information. Furthermore, as per References 5 through 7, there is some evidence that increments in workload have detrimental impacts to a certain degree on offshore workers' psychological well-being.

5.3.1 Offshore Worker Situation Awareness-Related Studies and Their Results

Over the years, a number of studies concerning offshore workers' situation awareness have been performed. Two such studies and their results are presented below, separately.

5.3.1.1 Study I

This study consists of interviews with personnel involved with offshore drilling, and the aim of the study was to understand how well the situation awareness concept is recognized within the offshore industrial sector. During the interviews, the following six questions were asked [5]:

1. How is situation awareness known in the offshore industry?
2. What factors affect the quality of a person's awareness?
3. What are the indicators of reduced awareness?
4. How can reduced awareness be improved?
5. How is team situation awareness achieved?
6. What can be done to check the awareness of workers?

The question "How is situation awareness known in the offshore industry?" received three responses. They were safety accountability, positional awareness, and safety awareness. The question "What factors affect the quality of a person's awareness?" received the following 12 responses [3,5]:

1. Fatigue
2. Routine task/complacency
3. Stress and workload
4. Conflict
5. Home/family problems
6. Weather/seasons
7. Experience (and new personnel)
8. Job prospects
9. Communication (good and bad)
10. Daydreaming
11. Having a near miss
12. Supervisory responsibility

The question "What are the indicators of reduced awareness?" received five responses. They were repetition of instructions, character change, reduction in communication, an expressionless appearance, and reduced work standards.

The question "How can reduced awareness be improved?" received 10 responses. They were increased involvement in rig activities, discussion of events, alter the work level, communication, removal from the situation, training, placing them (personnel) in a different job, problem solving, interaction, and alter the crew lineup. The question "How is team situation awareness achieved?" received the following eight responses [3,5]:

1. Adaptability
2. Increased interaction
3. Experience

4. Trust

5. Understand capabilities and traits planning

6. Consistency

7. Cooperation

8. Time

Finally, the question "What can be done to check the awareness of workers" received four responses. Two of these responses were constant assessment of surroundings and risk.

Additional information on the preceding responses is available in Reference 5.

5.3.1.2 Study II

This study was concerned with situation awareness errors in offshore drilling incidents. The study reported three categories of such errors along with their occurrence percentages as shown in Figure 5.2 [5].

The breakdown of the perception-related incident errors along with their occurrence percentages were as follows [3,5]:

- Failure to monitor or observe data: 26.8%
- Hard to discriminate or detect data: 15.7%
- Misconception of data: 14.2%
- Data not available: 9.7%
- Memory loss: 0.1%

Similarly, the breakdown of comprehension-related incident errors along with their occurrence percentages were use of incorrect mental model: 11.1%,

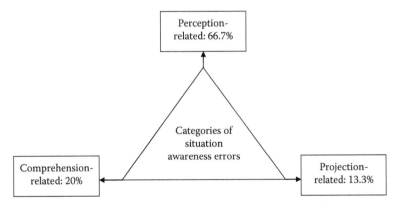

FIGURE 5.2
Categories of situation awareness errors in offshore drilling incidents and their corresponding occurrence percentages.

lack of/poor mental model: 6.7%, and overreliance on default values: 2.2%. Finally, the breakdown of projection-related incident errors along with their occurrence percentages were lack of/poor mental model: 13.2% and over projection of current trends: 0.1%.

5.4 Offshore Industry Accident Reporting Approach and Offshore Accident-Related Causes

Usually, the accident reporting approach used in the offshore industrial sector can be divided into the following two areas [3,8]:

1. *Minor accidents*: In this case, the investigators are safety officers and supervisors who normally have some training in the area. As the need arises, an investigation team is formed to conduct a more comprehensive investigation into the accident occurrence.
2. *Serious incidents or accidents*: In this case, investigation teams along with government-appointed accident inspectors fly from onshore to the offshore site. Normally, all incidents/accidents are processed and documented through specified channels and the accident reports' final copies are distributed to all the concerned authorities and bodies.

Usually the forms used for accident reporting contain information on items such as follows [3,8]:

- Time, date, and location of occurrence.
- Immediate and underlying causes.
- Equipment being used, including safety devices and equipment
- Equipment failures.
- The type of work being performed and experience of all involved individuals.
- All permits being issued and procedures being followed.
- Other people performing their tasks in the surrounding area.
- Protective clothing being worn by all involved personnel.
- Contributory factors (e.g., environmental conditions, any existing hazards).
- Personal details of all personnel involved, including supervisor at the time of the incident/accident occurrence.
- Type of accident/incident: property damage, material loss, injury, process disruption, flammable or poisonous substance leaks, fire

or explosion, dangerous occurrences, hazards, disease, near misses, and environmental harm.

5.4.1 Offshore Accident-Related Causes

A study of the accident reporting forms used by 25 offshore companies in the United Kingdom reported that there were a large number of immediate causes for the occurrence of accidents [8]. Most of these immediate causes were as follows [3,8]:

- Use of defective equipment
- Equipment used improperly
- Proper equipment not used
- Adjusted equipment in operation
- Serviced equipment in operation
- Operating without proper authority
- Work carried out on live or unsafe equipment
- Wrong speed
- Improper loading/lifting
- Workers under the influence of alcohol/drugs
- Lack of attention/forgetfulness
- Failure to warn/secure
- Safety device/equipment made inoperable

In addition, there were also a large number of underlying causes for the occurrence of accidents [8]. These underlying causes are grouped under two categories as shown in Figure 5.3 [3,8]. The two categories shown in Figure 5.3 are job factors and personal factors.

The job factors category has the following four elements [3,8]:

1. *Organization*: Its subelements were poor safety plan, company policy, poor procedures, competence standards, working hour policies, adequacies of systems, poor staffing and resources, and safety systems.

FIGURE 5.3
Categories of underlying causes for the occurrence of offshore accidents.

2. *Task*: Its subelements were inappropriate matching of individual to job task, failure in communication, inadequate or no job description, poor equipment selection, conflicting goals, inadequate work planning, time problems, and confusing directions.

3. *Management*: Its subelements were management practices, poor planning, bad examples set by management, management job knowledge, communication, and qualification and experience criteria.

4. *Supervision*: Its subelements were insufficient supervisory job knowledge, poor supervisory examples, incomplete instruction and training, poor work planning, improper production incentives, unclear responsibilities, inadequate discipline, and lack of inspections.

Similarly, the personal factors category has the following four elements [3,8]:

1. *Capability*: Its subelements were lack of competence, poor judgment, lack of mental capability, memory failure, perception demands, lack of physical capability, concentration demands, inability to comprehend, and judgment demands.

2. *Knowledge and skill*: Its subelements were lack of education, lack of experience, misunderstood directions, poor training, lack of job instructions, poor orientation, lack of awareness, inadequate practice, and lack of hands-on instructions.

3. *Stress*: Its subelements were general stress, fatigue, health hazards, frustration, and monotony.

4. *Improper motivation*: Its subelements were peer pressure, recklessness, lack of anticipation, inattention, inappropriate attempt to save time, attitude, aggression, horseplay, and inadequate thought and care.

Additional information on all of the above elements is available in References 8 and 9.

5.5 Offshore Industry Accidents' Case Studies

Over the years, a large number of accidents in offshore industry have occurred. Nine of the deadliest of these accidents are described below, separately.

5.5.1 Mumbai High North Platform Accident

The Mumbai high field, discovered in 1974, is the largest oil and gas field in India. The field is located in the Arabian Sea, approximately 100 miles west of Mumbai coast. The Mumbai High North Platform was constructed in 1981 and was an oil and natural gas processing complex owned and operated by India's state-owned Oil and Natural Gas Corporation [10].

The complex produced around 120,000 barrels of oil and approximately 4.4 million cubic meters of gas per day [10]. The platform was a seven-storey high steel structure, and it contained five gas export risers and 10 fluid import risers that were located just outside its jacket. On July 25, 2005, a multipurpose support vessel (MSV) named "Samudra Surakasha" collided with the platform that caused rupture of one or more of the platform's gas export risers. The reluctant gas leakage resulted in ignition that set the platform on fire, and the heat radiation caused damage to the Noble Charlie Yester jack-up rig engaged in drilling operation close to the platform. Furthermore, heat radiation also caused damage to MSV.

All in all, the platform was destroyed within hours and caused 22 deaths. Additional information on this accident is available in References 11 and 12.

5.5.2 Piper Alpha Accident

This accident is considered as the deadliest offshore oil rig accident in history. Piper Alpha was a North Sea oil production platform, and it was located approximately 120 miles northeast of Aberdeen, United Kingdom. The platform became operational in 1976 and was operated by Occidental Petroleum (Caledonia) Ltd. Initially, the platform was constructed to produce crude oil, but later on with the installation of gas conversion equipment, it also started producing gas.

Piper Alpha produced oil and gas from its 24 wells for delivery to the Flotta oil terminal located on the Orkney Islands as well as to other installations through three separate pipelines. At the time of the occurrence of the disaster, the platform, as per References 13 through 16, produced about 10% of North Sea oil and gas. On July 6, 1988, due to gas leakage from one of the condensate pipes at the platform, explosions and a resulting fire destroyed the platform and killed 167 persons [14,15].

A subsequent investigation into the disaster was conducted by the United Kingdom government that identified a number of factors that, directly or indirectly, contributed to the Piper Alpha incident's severity. Two of these factors were as follows [13–15]:

1. Serious breakdown in the chain of command and lack of any proper communication to the platform's crew members.
2. Existence of fire walls and lack of blast walls. More clearly, the existing fire walls predated the gas conversion equipment's installation as well as were not adequately upgraded to blast walls subsequent to the installation.

The investigation made 106 recommendations for changes to existing North Sea safety-related procedure. The offshore industry accepted all the recommendations. Additional information on Piper Alpha accident is available in References 10 and 13 through 16.

5.5.3 Bohai 2 Oil Rig Accident

The Bohai No. 2 oil rig was located in the Gulf of Bohai off the coast of China, and it was managed and operated by the Ocean Oil Company, China Petroleum Department. The rig was a self-elevating drilling unit that sank on November 25, 1979 while being towed after encountering a storm with force 10 winds [3]. The incident caused the death of 72 out of 76 persons on board the rig.

The postdisaster investigations into the accident attributed many causes for its severity. Four of these causes are presented below [3,17].

1. Poor training of the members of the crew in regard to the use of lifesaving-related equipment.
2. Failure to properly stow deck equipment before towing.
3. Poor emergency evacuation-related procedures.
4. Failure to properly follow standard tow-related procedures in regard to weather.

Additional information on this accident is available in References 10 and 17 through 19.

5.5.4 Alexander L. Kielland Accident

Alexander L. Kielland was a Norwegian semisubmersible rig/platform in the Ekofisk oil field, Norwegian continental shelf, about 235 miles east of Dundee, Scotland, United Kingdom. The rig/platform was named after a Norwegian writer, and it was owned by the Stavanger Drilling Company of Norway. At the time of the occurrence of the disaster, the rig/platform was hired by the United States Company called Phillips Petroleum.

After about 40 months of service, the rig/platform was no longer used for drilling, but it served as a so-called flotel (i.e., a floating hotel) for workers from the close by Edda platform. On March 27, 1980, wind gusts of approximately 40 knots created waves up to 12 m high that, in turn, caused the rig/platform to collapse into the North Sea, and resulted in the death of 123 off-duty workers.

A subsequent investigation conducted by the Norwegian government concluded that the rig collapsed due to a fatigue crack in one of the rig's six bracings (bracing D-6), which connected the collapsed D-leg to rest of the rig [3,20]. Additional information on this accident is available in References 3, 10, and 20.

5.5.5 Enchova Central Platform Accident

The Enchova Central Platform was located in the Campos Basin close to Rio de Janeiro, Brazil, and it was operated by the Brazilian company called Petrobras. On August 16, 1984, the accident occurred due to a blowout, which, in turn, caused fire and explosion at the central platform. Although most of the workers were evacuated safely from the platform by helicopter and life-boats, 42 of them lost their lives during the evacuation process [3,10].

More clearly, in this case, the most serious incident occurred when the lowering mechanism of a lifeboat malfunctioned. Consequently, the lifeboat remained vertically suspended until the stern support broke and the life-boat fell approximately 20 m deep into the sea and killed 36 of its occupants. Another six people got killed as they jumped approximately 40 m from the platform into the sea.

Additional information on this accident is available in References 10 and 21.

5.5.6 Ocean Ranger Accident

The Ocean Ranger was a semisubmersible mobile offshore drilling unit built by Mitsubishi Heavy Industry's Yard in Hiroshima, Japan, in 1976, for a Canadian company, Ocean Drilling and Exploration Company (ODECO). The drilling rig contained a square upper hull supported by eight vertical columns. The Ocean Ranger sank with 84 crew members on board during drilling an exploration well approximately 166 miles east of St. John's, Newfoundland, Canada on February 15, 1983.

A subsequent investigation into the disaster was conducted by a Canadian Royal Commission. The commission concluded that the rig had some design and construction-related flaws, particularly in the blast control room [22]. Furthermore, the Royal Commission also pointed out that the members of the crew lacked appropriate equipment, survival suits, and training. Additional information on this accident is available in References 10, 22, and 23.

5.5.7 Glomar Java Sea Drillship Accident

The Glomar Java Sea Drillship was built in 1975 to drill wells down to about 25,000 ft in water depths of approximately up to 1000 ft. The drill-ship was designed by Global Marine, Inc., and constructed by the Livingston Shipbuilding company of Orange, Texas. The 40-ft-long drillship was con-tracted to ARCO China, and it arrived in the South China Sea in January 1983 [10].

On October 25, 1983, the drillship capsized and sank in the South China Sea, at the depth of 317 ft, approximately 63 nautical miles south west of Hainan Island, China and about 80 nautical miles east of the Socialist Republic of Vietnam. The incident resulted in the death of 81 persons on board the drillship.

A subsequent investigation conducted by the United States National Transportation Safety Board concluded that the most likely cause for the Glomar Java Sea Drillship's capsizing and sinking was the flooding of its starboard tanks 6 and 7 through a hull fracture during typhoon Lex [24].

Additional information on this accident is available in References 10 and 24.

5.5.8 Baker Drilling Barge Accident

The Baker Drilling Barge accident occurred on June 30, 1964 in the Gulf of Mexico. Fire and an explosion on the drilling barge resulted in 21 deaths and 22 injuries [10]. At the time of the accident, the Baker Drilling Barge was deployed for drilling operation for Pan American Petroleum Corporation in Eugene Island, Gulf of Mexico.

In the morning of June 30, 1964, the drilling barge's two 260-ft-long hulls suffered a blowout. Just minutes after the blowout, the entire drilling barge was engulfed with fire and explosion. After heeling the aft for about half hour, the vessel sank upside down in the water.

Additional information on this accident is available in Reference 10.

5.5.9 Seacrest Drillship Accident

The Seacrest Drillship accident occurred on November 3, 1989 in the South China Sea about 430 km south of Bangkok, Thailand. At the time of the accident, the drillship was anchored for drilling at the Platong gas field owned and managed by Unocal [10]. On the day of the accident, Typhoon Gay produced about 40 ft high waves that capsized the drillship.

The accident resulted in the death of 91 crew members. Although on November 4, 1989, the drillship was reported missing, it was only found floating upside down the following day by a search helicopter. It is believed that the drillship's capsize occurred so fast that there was no distress signal and no time left for the crew members to respond to the accident.

Additional information on this accident is available in Reference 10.

PROBLEMS

1. Write an essay on safety in offshore oil and gas industry.
2. Discuss offshore industry risk picture.
3. Define situation awareness and describe the offshore worker situation awareness concept.
4. Describe at least one offshore worker situation awareness-related study and its results.
5. What are the items on which offshore industry accident reporting forms usually contain information?
6. Discuss offshore accident-related causes.

7. Describe the Piper Alpha accident.

8. Compare the Piper Alpha accident with the Mumbai High North Platform accident.

9. Describe the following two accidents:
   - Alexander L. Kielland accident
   - Bohai 2 Oil Rig accident

10. What was the name of the company that constructed the Glomar Java Sea Drillship and how many deaths occurred in the following accidents?
    - Seacrest Drillship accident
    - Enchova Central Platform accident
    - Ocean Ranger accident
    - Piper Alpha accident

References

1. Offshore Drilling, retrieved on June 19, 2015 from website: https://en.wikipedia.org/wiki/Offshore_drilling (last modified on January 13, 2016).

2. About offshore Oil and Gas Industry, retrieved on June 19, 2015 from website: http://www.modec.com/about/industry/oil-gas.html.

3. Dhillon, B.S., *Mine Safety: A Modern Approach*, Springer-Verlag, London, 2010.

4. Tveit, O.J., Safety issues on offshore process installation: An overview, *Journal of Loss Prevention in the Process Industries*, 7(4), 1994, 267–272.

5. Sneddon, A., Mearns, K., Flin, R., Situation awareness and safety in offshore drill crews, *Cognition, Technology, and Work*, 8(4), 2006, 255–267.

6. Parkes, K., Psychosocial aspects of stress, health and safety on north sea installations, *Scandinavian Journal of Work, Environment, and Health*, 24(5), 1988, 321–333.

7. Sutherland, K., Flin, R., Stress at sea: A review of working conditions in the offshore oil and fishing industries, *Work Stress*, 3, 1989, 269–285.

8. Gordon, R.P.E., The contribution of human factors to accidents in the offshore oil industry, *Reliability Engineering and System Safety*, 61, 1998, 95–108.

9. Bird, F.E., Germain, G.L., *Practical Loss Control Leadership: The Conservation of People, Property, Process, and Profits*, Institute Publishing (ILCI), Longville, Georgia, 1989.

10. The World's Worst Offshore Oil Rig Disasters, retrieved on January 20, 2015 from website: http://www.offshore-technology.com/features/feature-the-worlds-deadliest-offshore-oil-rig.

11. Mumbai High North, retrieved on April 10, 2009 from website: http://home.versatel.nl/the-sims/rig/mhn.htm.

12. Riser Safety in UK Waters-Lessons from Mumbai High North Disaster, Report No. SPC/Technical/OSD/33, Hazardous Installations Directorate, Health and Safety Executive, London, May 2006.

13. Hulll, A.M., Alexander, D.A., Klein, S., Survivors of the piper alpha oil platform disaster: Long-term follow up study, *The British Journal of Psychiatry*, 181, 2002, 433–438.

14. Pate-Cornell, M.E., Learning from the Piper Alpha accident: Analysis of technical and organizational factors, *Risk Analysis*, 13(2), 1993, 215–232.

15. Pate-Cornell, M.E., Risk analysis and risk management for offshore platforms: Lessons from the Piper Alpha accident, *Journal of Offshore Mechanics and Arctic Engineering*, 115(1), 1993, 179–190.

16. Petrie, J.R., Piper Alpha Technical Investigation Interim Report, Petroleum Engineering Division, Department of Energy, London, UK, 1988.

17. Bohai 2 Jack-Up, Retrieved on April 10, 2009 from website: http://home. versatel.nl/the-sims/rig/bohai2.htm.

18. Santos, R.S., Feijo, L.P., Deepwater safety challenges to consider in a fast-paced development environment, *Offshore*, 68(3), 2008, 1–5.

19. Santos, R.S., Feijo, L.P., Safety challenges associated with deepwater concepts utilized in the offshore industry, *Proceedings of the 9th International Symposium on Maritime Health*, 2007, pp. 1–8.

20. Alexander, L., Kielland Accident, Report of a Norwegian Public Commission Appointed by Royal Decree of March 28, 1980, Report No. ISBN B0000ED27N, Norwegian Ministry of Justice Police, Oslo, Norway, March 1981.

21. One Hundred Largest Losses, Marsh Risk Consulting, retrieved on April 10, 2009 from website: http://www.marshrisk consulting.com/st/PSEV-C-352-NR-304-htm.

22. Report of the Royal Commission on the Ocean Ranger Marine Disaster, Report ISBN No. 0660116820, Government of Canada, Ottawa, 1984.

23. Mobile Offshore Drilling Unit (MODU) Ocean Ranger (O.N.615641) Capsizing and Sinking in the Atlantic Ocean, Report No. USCG 0001 HQS 82, United States Coast Guard, Department of Transportation, Washington, DC, 1983.

24. Capsizing and Sinking of the United States Drillship Glomar Java Sea, Report No. NTBS-MAR-84-08, National Transportation Safety Board, Washington, DC, 1984.

6

Case Studies of Oil Tanker Spill-Related Accidents and Oil Tanker Spill Analysis

6.1 Introduction

Billions of metric tons of oil is shipped every year by tankers around the globe, and many oil tanker spill-related accidents occur. For example, 2.42 billion metric tons of oil was shipped by tankers in 2005 [1] and according to the International Tanker Owners Pollution Federation (ITOPF) since 1984–2007, 9351 accidental spills occurred between 1984–2007. [1,2].

Most of these spills resulting from routine operations, such as discharging cargo, taking on fuel, and loading cargo, are small (i.e., less than 7 metric tons per spill). In contrast, spills resulting from accidents such as hull failures, collisions, explosions, and groundings are much larger and generally involve losses of over 700 metric tons per spill [1,2].

From 1970, ITOFF has analyzed oil tanker spill-related accidents data and since 1974, it has maintained a database of accidental oil spills from tankers, combined carriers, and barrages.

This chapter presents various important aspects of case studies of oil tanker spill-related accidents and analysis.

6.2 Case Studies of Oil Tanker Spill-Related Accidents

Over the years, many oil tanker spill-related accidents have occurred around the globe. The case studies of some of these accidents are presented below, separately.

6.2.1 Independenta Accident

This accident occurred on November 15, 1979 when a Romanian oil tanker named "Independenta" collided with the Greek cargo ship called

"Evriali" in Bosphorus strait in Turkey. As a result of the accident, a part of Independenta carrying 93,800 tons of Libyan Es Sider crude oil spilled and caught fire [3,4]. There were also 26 tons of heavy fuel oil bunkers onboard the tanker.

In the ensuing explosion and fire onboard the tanker as well as from burning oil on the water, 42 crew members got killed. In addition, some buildings up to 6 km away were also damaged. On November 18 and December 6, more explosions occurred on the vessel that caused more release of oil to the sea. Most of the oil was consumed in the fire, with only small contamination of close-by shorelines. For a number of weeks, the Bosphorus strait was closed to traffic.

Additional information on this accident is available in References 3 and 4.

6.2.2 Sea Star Accident

This accident occurred on December 19, 1972 when a South Korean oil tanker named "Sea Star" collided with the Brazilian oil tanker called "Horta Barbosa" in the Gulf of Oman [3,5,6]. Sea Star was on voyage from Ras Tanura, Saudi Arabia to Rio de Janeiro, Brazil. Both tankers caught fire but while the fire on Horta Barbosa was extinguished within a day, Sea Star continued to burn and experienced a number of explosions.

Five days after the collision, Sea Star sank. The accident resulted in the death of 12 crew members and the spillage of around 115,000 tons of crude oil [5].

Additional information on this accident is available in References 3, 5, and 6.

6.2.3 Haven Accident

This accident occurred on April 11, 1991 when an oil tanker named "Haven" suffered an explosion while anchored in Genoa Roads, Italy. The vessel was loaded with 144,000 tons of Iranian heavy crude oil, and the explosion was caused by an electrical spark during tank cleaning [3,7,8]. The tanker broke into three sections and six crew members got killed.

As per Reference 6, approximately half of the crude oil onboard was burnt during the explosion and fire. Furthermore, roughly 10,000 tons of oil was spilled prior to the sinking of the main section of the tanker.

Additional information on this accident is available in References 3, 7, and 8.

6.2.4 ABT Summer Accident

This accident occurred on May 28, 1991 when an oil tanker named "ABT Summer" experienced an explosion and fire approximately 900 miles off the coast of Angola. The vessel was carrying a cargo of 260,000 tons of Iranian heavy crude oil. Five of the 32 crew members onboard were killed as a result of the accident [3,9].

A slick covering an area of about 80 square miles spread around the oil tanker and burnt quite fiercely. The vessel burnt continuously for three days and sank on June 1, 1991. The subsequent efforts for locating the ship wreckage were unsuccessful.

Additional information on this accident is available in References 3 and 9.

6.2.5 Jakob Maersk Accident

This accident occurred on January 29, 1975 when a Danish oil tanker named "Jacob Maersk" struck bottom during maneuvering onto the berth using tugs at a port in Portugal. The tanker was carrying approximately 88,000 tons of Iranian light crude oil and bunker fuel oil at the time of accident [3,10,11]. Heavy swell was the cause for the tanker to rise and fall continuously onto the rocky seabed that resulted in damage to the cargo tanks. Consequently, oil entered the engine room of the vessel that caused explosion and fire, which ultimately spread to the ship cargo holds as well as to the floating oil on the water.

The accident resulted in the death of seven tanker crew members, and the tanker and floating oil continued to burn for two days [11]. The fire consumed between 40,000 and 50,000 tons of oil and most of the spilled oil was blown out to sea.

Additional information on this accident is available in References 3, 10, and 11.

6.2.6 Hawaiian Patriot Accident

This accident occurred on February 23, 1977 when an oil tanker named "Hawaiian Patriot" reported a crack in its hull plating during a storm that resulted in a leak of oil from its cargo of 99,000 tons of light Indonesian crude oil. The tanker was approximately 300 miles west of Hawaii and about 18,000 tons of its oil leaked into the sea and on the following day (i.e., on February 24, 1977), the vessel caught fire and exploded [3,5,12].

For several hours, the tanker burnt fiercely and then sank with the remaining cargo onboard. The accident caused the death of one of the crew members.

The resultant oil slick that contained approximately 50,000 tons of oil was carried westward from Hawaii by ocean currents. Thus, there was no pollution problem on land. Additional information on this accident is available in References 3, 5, and 12.

6.2.7 Torrey Canyon Accident

This accident occurred on March 18, 1967 when an oil tanker named "Torrey Canyon" carrying a cargo of approximately 119,000 tons of Kuwait crude oil ran aground on Pollard Rock on the Seven Stones Reef, Cornwall, England [3,13–15]. The ruptured tanks of the vessel started to spill oil and within the next 12 days, the entire oil cargo of the tanker was lost.

The spilled oil from the vessel polluted many parts of the South West of England that caused the death of thousands of seabirds, and threatened the livelihoods of many people in the area in the forthcoming summer tourist season [15]. Later on, the drifting oil polluted beaches and harbors in the Channel Islands and Brittany areas of England.

Additional information on this accident is available in References 3 and 13 through 15.

6.2.8 Exxon Valdez Accident

This accident occurred on March 24, 1989 when an oil tanker named "Exxon Valdez" was grounded on Bligh Reef in Prince William Sound, Alaska. This resulted in the escape of approximately 37,000 tons of Alaska North Slope crude oil into the Prince William Sound and spread widely [3,16–18]. Despite the use of a large number of vessels, skimmers, and booms, only less than 10% of the original oil spill volume was possible to recover from the sea surface.

For the first year alone, the cost of cleanup was estimated to be more than $2 billion. As per Reference 18, Exxon Mobil paid $4.3 billion as a consequence of the spill that included cleanup costs; the costs of various legal settlements, criminal fines, and court verdicts; etc.

As a result of the accident, around 1000 sea otters died and more than 35,000 dead birds were retrieved. Additional information on this accident is available in References 3 and 16 through 18.

6.2.9 Irenes Serenade Accident

This accident occurred on February 23, 1980 when a Greek oil tanker named "Irenes Serenade," while at anchor at the bunkering location in Navarino Bay, Greece, suffered explosions in the forecastle that set its cargo alight. The vessel was loaded with a cargo of 102,660 tons of Iraqi crude oil [3,19]. As a result of the accident, an oil slick 2 miles long by half mile wide spread from the tanker and for the next 14 h both the vessel and the surrounding water burnt, and then the tanker sank.

All but two crew members of the ship were rescued and between 10 and 20 thousand tons of oil was observed in the open sea, two days after the tanker sank. Additional information on this accident is available in References 3 and 19.

6.2.10 Urquiola Accident

This accident occurred on May 12, 1976 when an oil tanker named "Urquiola" struck bottom on entering the port of La Coruna in Spain and started to leak oil. The vessel was carrying Saudi Arabian light crude oil. In order to avoid the risk of an explosion within the harbor area, the decision was made that the tanker should return to sea where repairs or offloading could take place safely [3,20,21].

Thus, the ship on its way out struck its bottom again and then ran hard aground between the two entrance channels. Prior to several explosions and fire, most of the crew members abandoned the vessel, but its master got killed.

Approximately 100,000 tons of crude oil was spilt during the incident, most of which burnt, but between 25 and 30 thousand tons washed ashore. Additional information on this accident is available in References 3, 20, and 21.

6.2.11 Hebei Spirit Accident

This accident occurred on December 7, 2007 when an oil tanker named "Hebei Spirit" was struck by a crane barge while at anchor off Taean, South Korea. The barge broke free from its tow because of bad weather and then it punctured three port-side cargo tanks of the vessel. The tanker was carrying 209,000 tons of four different Middle Eastern crude oils and due to the accident, approximately 10,900 tons of crude oil was released to the sea [22].

The spill's impact extended across three provinces of South Korea and several hundred miles of coastline, both on the mainland and on several islands, along South Korea's western coast. In the terms of staff involvement of ITOPF, this accident is considered to be the largest incident.

Additional information on this accident is available in Reference 22.

6.2.12 Atlantic Empress Accident

This accident occurred on July 19, 1979 when an oil tanker named "Atlantic Empress" collided with a vessel called "Agean Captain" about 10 miles off Tobago island during a tropical rainstorm. Immediately after the collision, both tankers started to leak oil and both caught fire. This resulted in the death of several crew members [3,23].

The burning vessel (i.e., Atlantic Empress), on July 21 and 22, was towed further out to sea. Seven days later, a large explosion caused severe damage to the tanker, and it sank on August 2, 1979. As per Reference 23, approximately 287,000 tons of oil was spilled from the vessel, which is considered to be the largest ship-source spill ever documented.

Additional information on this accident is available in References 3 and 23.

6.2.13 Castillo de Bellver Accident

This accident occurred on August 6, 1983 when an oil tanker named "Castillo de Bellver" carrying 252,000 tons of light crude oil caught fire approximately 70 miles northwest of Cape Town, South Africa [24–26]. The blazing vessel drifted offshore and broke into two sections, and its stern section, carrying around 100,000 tons of oil in its tanks, capsized and then sank into deep water about 24 miles from the coastline.

As a result of this accident, about 50,000–60,000 tons of crude oil was spilled into the sea or burnt. Although, this considerable amount of oil entered the sea, but there was a very little need for cleanup and its environmental effects were quite minimal.

Additional information on this accident is available in References 24 through 26.

6.2.14 Amoco Cadiz Accident

This accident occurred on March 16, 1978 when an oil tanker named "Amoco Cadiz" carrying 223,000 tons of light Iranian and Arabian crude oil ran aground off the coast of Brittany, France. Approximately 4000 tons of bunker fuel was released into heavy sea, and most of it formed a viscous water-in-oil emulsion that increased the volume of pollutant by up to five times [27–29]. Oil and emulsion contaminated approximately 320 km of the coastline of Brittany, France by the end of April and extended as far east to Channel Island, UK.

The accident resulted in the greatest loss of marine life ever recorded after the occurrence of an oil spill as millions of dead mollusks, sea urchins, and other benthic species washed ashore two weeks after the accident [29]. Furthermore, about 20,000 dead birds were recovered, in which diving birds constituted the majority.

Additional information on this accident is available in References 27 through 29.

6.2.15 Odyssey Accident

This accident occurred on November 10, 1988 when a Liberian oil tanker named "Odyssey" carrying 132,157 tons of North Sea Brent crude oil broke into two sections and sank in heavy weather 700 miles off the coast of Nova Scotia, Canada [3,30]. As the vessel sank, fire started on its stern section and then the surrounding crude oil caught fire. Canadian Coast Guard was only able to come within approximately 1.75 miles of the burning vessel because of the rough weather conditions.

As this accident took place about 700 miles from the closest coastline, there were no concerns about pollution. Additional information on this accident is available in References 3 and 30.

6.2.16 Braer Accident

This accident occurred on January 5, 1993 when an oil tanker named "Braer" carrying 84,700 tons of Norwegian Gullfaks crude oil and about 1500 tons of heavy bunker oil, following its engine failure, ran aground in severe weather conditions on Garth's Ness, Shetland, UK [31]. Over a period of 12 days, the ship's entire cargo of oil was lost and the ship broke apart because of almost constant storm force winds and heavy seas.

In regard to the size of the spill, the oiling of shorelines was minimal. Furthermore, the oil spill of this ship was quite unusual in that significant amount of crude oil was spread to land adjacent to the ship's wreck site. Additional information on this accident is available in Reference 31.

6.2.17 Katina P Accident

This accident occurred on April 17, 1992 when an oil tanker named "Katina P" carrying 66,700 tons of heavy fuel oil from Venezuela to the United Arab Emirates was disabled by a freak wave during its transition through the Mozambique Channel. Approximately 3000 tons of heavy fuel oil was released into the channel because the vessel lost hull plating amid ships [3,32].

In order to prevent sinking, the tanker was intentionally grounded on a sandbar 6 miles offshore of Maputo Bay, Mozambique. During the towing process, the tanker broke into two and sank on April 26, 1992 and released more oil.

Additional information on this accident is available in References 3 and 32.

6.2.18 Prestige Accident

This accident occurred on November 13, 2002 when an oil tanker named "Prestige" carrying 77,000 tons of heavy fuel oil suffered hull damage in heavy seas off the northern part of Spain [33–35]. The vessel broke into two sections, about 170 miles west of Vigo, Spain, on November 19, 2002 and then sank in water 2 miles deep. Approximately 63,000 tons of heavy fuel oil was released into the sea from the tanker.

The released oil drifted for an extended period with winds and currents, and affected as far as the north coast of Spain and the Atlantic Coast of France. Approximately 141,000 tons of oil-related waste was collected in Spain and about 18,300 tons in France.

Additional information on this accident is available in References 33 through 35.

6.2.19 Sea Empress Accident

This accident occurred on February 15, 1996 when an oil tanker named "Sea Empress" carrying 130,000 tons of North Sea crude oil ran aground in the entrance to Milford Haven, Wales, UK. Although the vessel was refloated within few hours, it sustained serious damage to its center tanks and starboard that resulted in a massive oil release [36,37].

Severe weather conditions thwarted a number of attempts to bring the tanker under control and to undertake a ship-to-ship transfer operation.

The vessel released approximately 72,000 tons of crude oil and around 370 tons of heavy fuel oil into the sea.

Additional information on this accident is available in References 36 and 37.

6.2.20 Aegean Sea Accident

This accident occurred on December 3, 1992 when a Greek oil tanker named "Aegean Sea" ran aground during stormy weather while approaching the port of La Coruna, Spain. The vessel was carrying 80,000 tons of North Sea Brent crude oil and broke into two and caught fire which, along with spilled oil, burnt for a number of days [3,38,39].

The tanker's forward section sank in shallow water, about 50 m from the coastline and its stern section remained largely intact, and it still contained about 6500 tons of crude oil and approximately 1700 tons of bunker fuel. The accident resulted in the spillage of about 73,000 tons of oil, much of which was either consumed by fire onboard the tanker or dispersed at sea.

Additional information on this accident is available in References 3, 38, and 39.

6.2.21 Nova Accident

This accident occurred on December 6, 1985 when an oil tanker named "Nova" collided with an ultra large crude oil carrier named "Magnum" about 90 miles southeast of Khark Island in the Gulf of Iran [40]. Both the tankers were used as shuttle vessels for transporting oil between Khark Island and another island called "Sirri Island" in the area. The accident occurred due to the lack of running lights on both tankers.

The Nova tanker was carrying 188,000 tons of Iranian light crude oil and the collision damaged its five cargo tanks that resulted in a spill of around 70,000 tons of oil. Additional information on this accident is available in Reference 40.

6.2.22 Khark 5 Accident

This accident occurred on December 19, 1989 when an Iranian Oil tanker called "Khark 5" was damaged in a storm; it exploded and caught fire about 150 nautical miles off the cost of Morocco [41]. All crew members of the tanker were rescued by a passing cargo vessel. At the time of the accident, "Khark 5" was carrying 280,000 tons of Iranian heavy crude oil.

The accident damaged four cargo tanks of the tanker, resulting in a spill of about 70,000 tons of oil over a period of 12 days. The tanker was towed away, on January 1, 1990, from shore to the open ocean, and its remaining crude oil was trans-shipped into another tanker.

Additional information on this accident is available in Reference 41.

6.3 Tanker Oil Spills Number Analysis

Since 1970, over 1800 oil tanker spills, greater than 7 tons per spill, have occurred around the globe [42]. Tables 6.1 and 6.2 present the annual breakdowns of these spills under three categories: 7–700 tons spills, greater than 700 tons spills, total No. of spills of 7 tons and above for the periods 1970–1989 and 1990–2013, respectively [42].

Data presented in Tables 6.1 and 6.2 indicate that oil spills greater than 700 tons have decreased significantly over the period of 44 years. For example, for the period 1970–1979, the average number of such spills was 24.5 spills per year and for the period 2000–2009, the number was 3.5 spills per year.

Furthermore, 54% of these spills occurred during the period 1970–1979 and for the decade 2000–2009, this percentage decreased to 8%.

Also, from Tables 6.1 and 6.2, a decline in the occurrences of 7–700 tons spills can also be observed. For example, for the period 1990–1999, the average number of such spills was 28.1 spills per year, whereas for the period 2000–2009, it was 14.1 spills per year.

TABLE 6.1

Oil Tanker Spills of 7 Tons and Above for the Period 1970–1989

Year	No. of 7–700 Tons Oil Tanker Spills	No. of Greater than 700 Tons Oil Tanker Spills	Total No. of Oil Tanker Spills of 7 Tons and Above
1970	7	29	36
1971	18	14	32
1972	48	27	75
1973	28	31	59
1974	90	27	117
1975	96	20	116
1976	67	26	93
1977	70	16	86
1978	59	23	82
1979	60	32	92
1980	52	13	65
1981	54	7	61
1982	46	4	50
1983	52	13	65
1984	26	8	34
1985	33	8	41
1986	27	7	34
1987	27	11	38
1988	11	10	21
1989	32	13	45
Total	902	339	1242

TABLE 6.2

Oil Tanker Spills of 7 Tons and Above for the Period 1990–2013

Year	No. of 7–700 Tons Oil Tanker Spills	No. of Greater than 700 Tons Oil Tanker Spills	Total No. of Oil Tanker Spills of 7 Tons and Above
1990	50	14	64
1991	30	7	37
1992	31	10	41
1993	31	11	42
1994	26	9	35
1995	20	3	23
1996	20	3	23
1997	28	10	38
1998	25	5	30
1999	20	5	25
2000	21	4	25
2001	18	3	21
2002	11	3	14
2003	19	4	23
2004	19	5	24
2005	21	4	25
2006	12	5	17
2007	12	4	16
2008	7	1	8
2009	7	2	9
2010	4	4	8
2011	5	1	6
2012	7	0	7
2013	4	3	7
Total	448	120	568

6.4 Quantities of Oil Spilt

Since 1970, by considering oil tanker spills 7 tons and above, approximately 5.74 million tons of oil was lost as result of tanker incidents around the globe [42]. Tables 6.3 and 6.4 present the annual and accumulated quantities of oil spills of 7 tons and above since 1970 for the periods 1970–1989 and 1990–2013, respectively [42]. It is to be noted that the annual quantities presented in these two tables are rounded to the nearest thousand.

During the period 1990–1999, oil tanker spills of 7 tons and above resulted in 1,133,000 tons of oil lost, but around 73% of this amount was spilt in just 10 accidents/incidents [42]. Similarly, during the period 2000–2009, oil tanker spills of 7 tons and above resulted in 213,000 tons of oil lost, but 53% of this amount was spilt in just four accidents/incidents.

TABLE 6.3

Annual and Accumulated Quantities of Oil Tanker Spills of 7 Tons and Above for the Period 1970–1989

Year	Quantity of Oil Tanker Spills of 7 Tons and Above (Annual)	Accumulated Quantity of Oil Tanker Spills of 7 Tons and Above
1970	386,000	386,000
1971	144,000	530,000
1972	313,000	843,000
1973	159,000	1,002,000
1974	173,000	1,175,000
1975	351,000	1,152,600
1976	364,000	1,890,000
1977	276,000	2,166,000
1978	393,000	2,559,000
1979	636,000	3,195,000
1980	206,000	3,401,000
1981	48,000	3,449,000
1982	12,000	3,461,000
1983	384,000	3,845,000
1984	29,000	3,874,000
1985	85,000	3,959,000
1986	19,000	3,978,000
1987	38,000	4,016,000
1988	190,000	4,206,000
1989	164,000	4,370,000

6.5 Oil Spill Causes

Since 1970, around the world, there have been many different causes for the occurrence of oil tanker spills. For the period 1970–2013, the percentages of such spill incidences greater than 700 tons by cause are presented in Table 6.5 [42].

Similarly, for the period 1970–2013, the percentages of spill incidences between 7 and 700 tons by cause are presented in Table 6.6 [42].

Finally, for the period 1974–2013, the percentages of spill incidences less than 7 tons by cause are presented in Table 6.7 [42].

It is interesting to note from Tables 6.5 and 6.6 that two highest percentages of causes for the spill incidences greater than 700 tons and for the spills between 7 and 700 tons are grounding and allision/collision. In contrast, the percentages of these two causes for the spills less than 7 tons are on the lower end (i.e., grounding: 3% and allision/collision: 2%).

Additional information on the oil spill causes is available in Reference 42.

TABLE 6.4

Annual and Accumulated Quantities of Oil Tanker Spills of 7 Tons and Above for the Period 1990–2013

Year	Quantity of Oil Tanker Spills of 7 Tons and Above (Annual)	Accumulated Quantity of Oil Tanker Spills of 7 Tons and Above
1990	61,000	4,431,000
1991	431,000	4,862,000
1992	167,000	5,029,000
1993	140,000	5,169,000
1994	130,000	5,299,000
1995	12,000	5,311,000
1996	80,000	5,391,000
1997	72,000	5,463,000
1998	13,000	5,476,000
1999	28,000	5,504,000
2000	14,000	5,518,000
2001	9,000	5,527,000
2002	66,000	5,593,000
2003	44,000	5,637,000
2004	16,000	5,653,000
2005	18,000	5,671,000
2006	23,000	5,694,000
2007	19,000	5,713,000
2008	2000	5,715,000
2009	3000	5,718,000
2010	12,000	5,730,000
2011	2000	5,732,000
2012	1000	5,733,000
2013	7000	5,740,000

TABLE 6.5

Percentages of Spill Incidences Greater than 700 Tons by Cause for the Period 1970–2013

No.	Spill Cause	Percentage (%)
1	Grounding	33
2	Allision/collision	30
3	Hull failure	13
4	Fire/explosion	11
5	Equipment failure	4
6	Other/unknown	9

TABLE 6.6

Percentages of Spill Incidences between 7 and
700 Tons by Cause for the Period 1970–2013

No.	Spill Cause	Percentage (%)
1	Allision/collision	26
2	Grounding	20
3	Equipment failure	15
4	Hull failure	7
5	Fire/explosion	4
6	Other/unknown	28

TABLE 6.7

Percentages of Spill Incidences Less than 7 Tons
by Cause for the Period 1974–2013

No.	Spill Cause	Percentage (%)
1	Equipment failure	21
2	Hull failure	7
3	Grounding	3
4	Fire/explosion	2
5	Allision/collision	2
6	Other/unknown	65

PROBLEMS

1. Describe the following two oil tanker accidents:
   - Independenta accident
   - Odyssey accident
2. Compare the following two oil tanker accidents:
   - Hawaiian Patriot accident
   - Amoco Cadiz accident
3. How many oil tanker spills of 7 tons and above have occurred since 1970–2013 around the globe?
4. Describe the following three oil tanker accidents:
   - Prestige accident
   - Sea Empress accident
   - Aegean Sea accident
5. How many 7–700 tons oil tanker spills occurred in 1977, 1979, 1999, and 2013?
6. Compare the following two oil tanker accidents:
   - Castillo de Bellver accident
   - Braer accident

7. What were the quantities of oil tanker spills of 7 tons and above in 1979, 1991, 2000, and 2013?
8. What were the percentages of incidences of oil tanker spills greater than 700 tons for the following causes: hull failure, grounding, and equipment failure?
9. Where and when did the Torrey Canyon and Exxon Valdez accidents occurred?
10. Write an essay on oil tanker spill-related accidents.

References

1. Oil Tanker, retrieved on June 9, 2015 from website: https://en.wikipedia.org/wiki/Oil_tanker (last modified on January 15, 2016).
2. Oil Tanker Spill Information Pack, retrieved on October 8, 2008 from website: http://www.itopf.com/information-sevices/data-and-statistics/statistics/. International Tanker Owners Pollution Federation, London.
3. Hooke, N., *Maritime Casualties, 1963–1996*, LLP Limited, London, 1997.
4. Independenta, Bosphorus, Turkey, 1979, retrieved on January 23, 2015 from website: http://www.itopf.com/in-action/case-studies/case-study/independenta-boshporus-turkey-1.
5. Butler, J.N., The Largest oil spills: Inconsistencies, information gaps; a review of 30 of the largest oil spills from 1942–1978, *Ocean Industry*, 13(10), 1978, 101–112.
6. Sea Star, Gulf of Oman, 1972, retrieved on January 23, 2015 from website: http://www.itopf.com/in-action/case-studies/case-study/sea-star-gulf-of-oman-1972/.
7. Martinelli, M. et al., The M/C haven oil spill: Environmental assessment of exposure pathways and resource injury, *Proceedings of the International Oil Spill Conference*, 1995, pp. 10–15.
8. Haven, Italy, 1991, retrieved on January 23, 2015 from website: http://www.itopf.com/in-action/case-studies/case-study/haven-italy-1991/.
9. ABT Summer, off Angola, 1991, retrieved on January 23, 2015 from website: http://www.itopf.com/in-action/case-studies/case-study/abt-summer-off-angola-1991/.
10. Canelas, L.D., Calejo Monteiro, J.D., Some studies of an oil spillage due to the Jacob Maersk Accident, *Proceedings of the Oil Spill Conference*, 1977, pp. 281–288.
11. Jacob Maersk, Leixoes, Portugal, 1975, retrieved on January 23, 2015 from website: http://www.itopf.com/in-action/case-studies/case-study/jacob-maersk-leixoes-portugal-1975/.
12. Hawaiian Patriot, off Hawaii, 1977, retrieved on January 23, 2015 from website: http://www.itopf.com/in-action/case-studies/case-study/hawaiian-patriot-off-hawaii-1977/.
13. Petrow, R., *The Black Tide: In the Wake of Torrey Canyon*, Hodder & Stoughton, London, 1968.
14. Southward, E.C., Southward, A.J., Recolonization of rocky shores in Cornwall after the use of toxic dispersants to clean up the Torrey canyon spill, *Journal of the Fisheries Research Board of Canada*, 35(5), 1978, 682–706.

15. Torrey Canyon, United Kingdom, 1967, retrieved on January 23, 2015 from website: http://www.itopf.com/in-action/case-studies/case-study/torrey-canyon-united-kingdom-1967/.

16. Loughlin, T.R., Editor, *Marine Mammals and the Exxon Valdez*, Academic Press, San Diego, California, 1994.

17. Galt, J.A., Lehr, W.J., Payton, D.L., Fate and transport of the Oxxon Valdez Oil Spill, *Environmental Science and Technology*, 25(2), 1991, 202–209.

18. Exxon Valdez, Alaska, United States, 1989, retrieved on January 23, 2015 from website: http://www.itopf.com/in-action/case-studies/case-study/exxon-valdez-alaska-united-states-1989/.

19. Irenes Serenade, Greece, 1980, retrieved on January 23, 2015 from website: http://www.itopf.com/in-action/case-studies/case-study/irenes-serenade-greece-1980/.

20. Gundlach, E.R., Hayes, M.O., The Urquiola Oil Spill, La Coruna, Spain: Case history and discussion of methods of control and clean-up, *Marine Pollution Bulletin*, 8, 1977, 132–136.

21. Urquilo, Spain, 1976, retrieved on January 23, 2015 from website: http://www.itopf.com/in-action/case-studies/case-study/urquinola-apin-1976/.

22. Hebei Spirit, Republic of Korea, 2007, retrieved on January 23, 2015 from website: http://www.itopf.com/in-action/case-studies/case-study/hebei-spirit-republic-of-korea-2007/.

23. Atlantic Empress, West Indies, 1979, retrieved on January 23, 1015 from website: http://www.itopf.com/in-action/case-studies/case-study/atlantic-empress-west-indies-1979/.

24. Wardley-Smith, J., The Castillo de Bellver, *Oil and Petrochemical Pollution*, 4(1), 1983, 291–293.

25. Moldan, A.G.S. et al., Some aspects of the Castillo de Bellver Oil Spill, *Marine Pollution Bulletin*, 16(3), 1985, 97–102.

26. Castillo de Bellver, South Africa, 1983, retrieved on January 23, 2015 from website: http://www.itopf.com/in-action/case-studies/case-study/castillo-de-bellver-south-africa-1983/.

27. Bellier, P., Massart, G., The Amoco Cadiz Oil Spill cleanup operations: An overview of the organization, control and evaluation of the cleanup techniques employed, *Proceedings of the Oil Spill Conference*, 1979, pp. 141–146.

28. Spooner, M.F., Editor, The Amoco Cadiz Oil Spill, *Special Edition of Marine Pollution Bulletin*, 9(7), 1978, 15–18.

29. Amoco Cadiz, France, 1978, retrieved on January 23, 2015 from website: http://www.itopf.com/in-action/case-studies/case-study/amoco-cadiz-france-1978/.

30. Odyssey, off Canada, 1988, retrieved on January 23, 2015 from website: http://www.itopf.com/in-action/case-studies/case-study/odyssey-off-canada-1988/.

31. Braer, UK, 1993, retrieved on January 23, 2015 from website: http://www.itopf.com/in-action/case-studies/case-study/braer-uk-1993/.

32. Katina, P, off Mozambique, 1992, retrieved on January 23, 2015 from website: http://www.itopf.com/in-action/case-studies/case-study/katina-p-off-mozambique-1992/.

33. Albaiges, J., Vilas, F., Morales-Nin, B., The prestige: A scientific response, *Marine Pollution Bulletin*, 53(5–7), 2006, 15–20.

34. Guillen, A.V., Prestige and the law: Regulations and compensation, *Proceedings of the Annual Conference on Oil Pollution*, 2004, pp. 50–54.

35. Prestige, Spain/France, 2002, retrieved on January 23, 2015 from website: http://www.itopf.com/in-action/case-studies/case-study/prestige-spainfrance-2002/.
36. Edwards, R., White, I., The Sea Empress Oil Spill: Environmental impact and recovery, *Proceedings of the International Oil Spill Conference*, 1999, pp. 97–102.
37. Sea Empress, Wales, UK, 1996, retrieved on January 23, 2015 from website: http://www.itopf.com/in-action/case-studies/case-study/sea-empress-milford-haven-wales-uk-1996/.
38. Prote, C. et al., The Aegean Sea Oil Spill one year after: Petroleum hydrocarbons and biochemical responses in Marine Bivalves, *Marine Environmental Research*, 42, 1996, 404–405.
39. Aegean Sea, Spain, 1992, retrieved on January 23, 2015 from website: http://www.itopf.com/in-action/case-studies/case-study/segan-sea-spain-1992/.
40. Nova, off Iran, 1985, retrieved on January 23, 2015 from website: http://www.itopf.com/in-action/case-studies/case-study/nova-off-iran-1985/.
41. Khark 5, off Morocco, 1989, retrieved on January 23, 2015 from website: http://www.itopf.com/in-action/case-studies/case-study/khark-5-off-morocco-1989/.
42. Oil Tanker Spill Statistics 2013, Report, The International Tanker Owners Pollution Federation (ITOPF) Limited, London, UK, 2014.

7

Human Factors Contribution to Accidents in the Oil and Gas Industry and Fatalities in the Industry

7.1 Introduction

In the past, accidents occurring in the industrial sector were reported generally in terms of technological failures, and the human element in the accident causation tended to be overlooked. Nowadays, the role of human factors has become more apparent, since the frequency of technology-related malfunctions has diminished quite considerably. Over the years, human factors and fatalities have become an important issue in the oil and gas industry.

For example, accidents such as the Piper Alpha disaster [1] clearly illustrate that a highly complex sociotechnological system performance is dependent upon the interaction of human, technical, organizational, social, environmental, and managerial elements. More clearly, all these factors or elements can be very important cocontributors to incidents that could result in catastrophic events. Thus, this chapter presents various important aspects of human factors contribution to accidents in the oil and gas industry and fatalities in the industry.

7.2 Human Factors That Affect Safety in General

Human factors that affect safety in general may be defined as organizational, group, and individual factors, as a study conducted by the Institute of Nuclear Power Operations (INPO) showed that the underlying causes for the occurrence of accidents in the nuclear industry covered organizational, group, and individual factors [2]. The underlying causes were divided into the following eight categories [3]:

1. Poor documentation or procedures (43%)
2. Lack of training or knowledge (18%)
3. Failure to follow proper procedures (16%)
4. Poor planning or scheduling (10%)
5. Miscommunication (6%)
6. Poor supervision (3%)
7. Policy-related problems (2%)
8. Others (2%)

The organizational, group, and individual factors are discussed below, separately.

7.2.1 Organizational Factors

The organizational factors that may contribute to incidents and accidents include cost-cutting programs and the degree of communication between work sites. More specifically, the factors that relate to safety are safety training, management commitment to safety, stable work force, positive safety promotion policy, environmental control and management, and open communication [3,4]. Furthermore, the factors that may discriminate between organizations in terms of safety climate are as follows [3,5]:

- Management attitudes toward safety
- Importance of safety training
- Status of safety committee
- Level of risk at the workplace
- Effect of safe conduct on social status
- Effect of safe conduct on promotion
- Effects of workplace
- Status of safety officer

As per References 6 and 7, in regard to the occurrence of the Piper Alpha disaster, the organizational and technical factors mainly stemmed from financial pressures. Consequently, some of the direct or indirect resulting factors were as follows [3]:

- Management personnel were under pressure to lower production-related costs. Thus, the costs of design, construction, inspection, and maintenance were at a minimal level.
- There was a "culture of denial" of the serious risks. This resulted in management focusing on frequent incidents that have the potential

to disrupt production rather than pinpointing its attention on the risk of a catastrophe.

- Incentives and rewards were given for short-term production figures. This could have encouraged all involved working personnel in cutting corners for getting the job finished on time or earlier.
- There was a high turnover of staff personnel. This may have resulted in poor understanding of the system, thus pushing the system to its limit.

All in all, four ways for organizations to learn from past experiences are as follows [3,8]:

1. Describe all accidents in safety bulletins and discuss them in safety meetings.
2. Ensure that all standards and codes of practice contain appropriate notes on accidents that led to the recommendations.
3. Make use of accident information retrieval and storage systems as they contain a lot of useful information.
4. Make compulsory reading, of a "black book" containing reports of accidents that have occurred to all newcomers.

7.2.2 Group Factors

Past experiences clearly indicate that the group factors play an important role in the safety of high-hazard industry, where communication between an organization's different members plays a major role. Other factors that contribute include relationship between individuals, leadership abilities of management personnel, and resources available to supervisors [3,9].

A study conducted by the U.S. National Transportation Safety Board reported that 73% of accidents were due to flight crew failures rather than technical-related problems [10]. Furthermore, the studies conducted by National Aeronautics and Space Administration (NASA) (in the form of pilot interviews, simulator observations, and accident analyses) reported that there was a definite need for more focus on team work and communication of pilots including command, decision making, and leadership [3]. Furthermore, the studies revealed that the crew factors such as the attitudes of the team toward communication and coordination, command responsibility, and recognition of stressor effects affect safety performance.

All in all, the postmortem analysis of the Piper Alpha disaster also highlighted the importance of crew factors to safety in the offshore oil industry [3,7].

7.2.3 Individual Factors

In the industrial sector, there are many individual factors that directly or indirectly affect safety. Some of these factors are as follows [3]:

- The clarity of the job instructions
- The degree/level of training and experience
- Not given enough responsibilities
- Being overworked

Additional information on individual factors is available in Reference 3.

7.3 Categorization of Accident-Related Human Factors in the Industrial Sector

Past experiences indicate that the categorization of accident-related human factors can vary from one type of industrial sector to another type. This sector presents such categorizations in two different industrial sectors.

In the area of power generation, a study of German nuclear power plants divided human factors affecting safety into the following eight categories [11]:

- *Category I: General aspects.* These include items such as time, affected parts, compound characteristics, and operational phases.
- *Category II: Organizational aspects.* These are cooperation between organizations and safety culture.
- *Category III: Job factors.* These include items such as task characteristics and content, procedures for task, information about task, and tools and safety devices.
- *Category IV: Personal aspects.* These are acting person characteristics and on-group characteristics.
- *Category V: Aspects of the failure.* These are the types of failure, trigger, and violations of procedures and rules.
- *Category VI: Aspects of causes.* These include items such as conditioning factors, erroneous decision making, and communication.
- *Category VII: Aspects of feedback.* These are error consequences, error discovery, and feedback characteristics.
- *Category VIII: External impacts.* These are flood and lightning.

In the area of aviation, Human Factors Reporting Program "BASIS" of the British Airways divided accident causation under the following five categories [3]:

- *Category I: Crew actions factors.* These are crew communication, assertiveness, decision process, briefing, planning, procedure, workload management, group climate, role conformity, as well as some human errors: memory lapse, mistake, action slip, misunderstanding, and misrecognition.

- *Category II: Organizational factors.* These are technical support and training, commercial pressure, group violation, company communication, maintenance, and recency-related problems.
- *Category III: Personal factors.* These are knowledge, tiredness, boredom, environmental awareness, environmental and operational stress, morale, recent practice, and distraction.
- *Category IV: Environmental factors.* These are unclear information, language, communication systems, operational problems, and weather conditions.
- *Category V: Informational factors.* These are manuals, electronic checklists, and standard operating procedures (SOPs).

7.4 Categories of Human Factors Accident Causation in the Oil Industry

A study of accident reporting forms of 25 offshore oil companies in the United Kingdom reported that the causes of accidents were grouped under two categories as shown in Figure 7.1 [3].

The "immediate causes" were either technical or human. There were a total of 15 causes. Thirteen of these causes were as follows [3]:

1. Improper speed
2. Improper lifting/loading
3. Operating without authority
4. Made use of defective equipment
5. Working on live or unsafe equipment

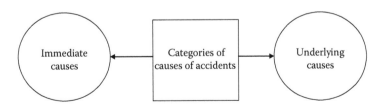

FIGURE 7.1
Categories of causes of accidents in the offshore oil industry.

6. Carried out adjustments to equipment in operation
7. Used equipment incorrectly
8. Lack of attention/forgetfulness
9. Under the influence of alcohol/drugs
10. Serviced equipment in operation
11. Made safety device inoperable
12. Did not use correct equipment
13. Failure to warn/secure

The "underlying causes" were grouped under "personal factors" and "job factors." There were four categories of the "personal factors" as shown in Figure 7.2 [3].

The elements of the "capability" are memory failure, physical capability, lack of competence, judgment demands, mental capability, perception demands, poor judgment, concentration demands, and inability to comprehend. The "improper motivation" elements are peer pressure, insufficient thought and care, inattention, recklessness, attitude, aggression, lack of anticipation, inappropriate attempt to save time, and horseplay.

The elements of the "stress" are fatigue, frustration, health hazards, monotony, and stress. Finally, the "knowledge and skill" elements are inadequate training, misunderstood direction, lack of experience, lack of education, lack of hands-on instructions, lack of job instructions, inadequate practice, lack of awareness, and poor orientation.

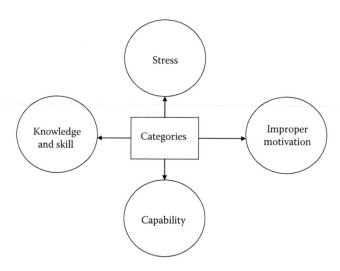

FIGURE 7.2
The "personal factors" categories.

The four categories of the "job factors" were as follows [3]:

1. Organization
2. Management
3. Task
4. Supervision

The elements of the "organization" are company policy, working hour policies, safety system, inadequate staffing and resources, poor safety plan, adequacies of systems, competence standards, and inadequate procedures. The "management" elements are poor planning, management job knowledge, bad management example, management practices, qualifications and experience criteria, and communication.

The elements of the "task" are work planning, poor or no job description, equipment selection, failure in communication, poor matching of individual to job task, confusing directions, time problem, and conflicting goals. Finally, the "supervision" elements are work planning, supervisory job knowledge, inspection, inadequate discipline, instruction training, supervisory example, improper production incentives, and unclear responsibilities.

7.5 Oilfield Fatality Analysis

A large number of fatalities occur every year in the oil and gas industry around the world. For example, in the United States alone, as per the Occupational Safety and Health Administration (OSHA) database, during the period 1997–2003, one fatality occurred every 10 days in the U.S. upstream oil and gas industrial sector [12]. Furthermore, as per the U.S. Bureau of Labor Statistics (BLS) report of 2003 data, in the oil and gas extraction subsector for every 100,000 workers, there were 34.5 fatalities or one fatality after every 4.3 days [12].

A study of the OSHA Integrated Management Information System (IMIS) database revealed that during the period 1997–2003, the U.S. exploration and production (E&P) industry experienced 254 fatalities onshore [12]. The analysis of the causes of the fatalities by incident type is presented in Table 7.1 [12].

It is to be noted from Table 7.1 that the main causes of fatalities as determined by the analysis were as shown in Figure 7.3.

Further analysis of the "struck by" incidents revealed that there was a relatively even split among the three discernible "struck by" causes. The "split" was as follows [12]:

1. Mechanical (39%)
2. Pressure (39%)
3. Gravity-dropped object (22%)

TABLE 7.1

Analysis of Oilfield Fatality by Incident Type

Incident Type No.	Incident Type Description	Percentage of Fatalities by Incident Type
1	"Struck by"	47
2	Explosions/fires	16
3	Falls from heights	14
4	Caught in between	7
5	Electrocution	6
6	Drowning	3
7	All others	7

The analysis of the causes of the fatalities by equipment type is presented in Table 7.2 [12].

It is to be noted from Table 7.2 that non-transportation-related vehicles were the equipment type associated with the most fatalities (i.e., 10%). Also, line pipe and tubular were major types of equipment involved with the oilfield-related fatalities.

All in all, the oilfield fatality analysis also revealed the following information:

- 33% of the fatalities occurred somewhere on the well site proper
- 20% of the fatalities occurred specifically on the rig floor
- 15% of the fatalities occurred in the derrick

Additional information on oilfield fatality analysis is available in Reference 12.

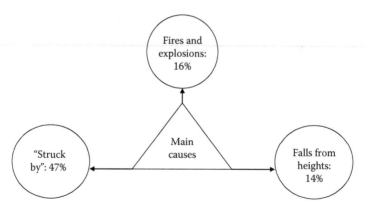

FIGURE 7.3
Main causes of fatalities as determined by the analysis.

TABLE 7.2

Analysis of Oilfield Fatality by Equipment Type

Equipment No.	Equipment Type Description	Percentage of Fatalities by Equipment Type
1	Vehicles	10
2	Line pipe	9
3	Tubulars	9
4	Draw works/hoists	8
5	Tank/vessels	7
6	Pump unit	6
7	Boards/baskets	5
8	Vac truck	5
9	Miscellaneous	18
10	Other	23

7.6 Recommendations to Reduce Fatal Oil and Gas Industry Incidents

The eight recommendations shown in Figure 7.4 could be very useful for reducing fatalities in oil and gas industry.

All of the recommendations shown in Figure 7.4 are described below, separately [12].

- *Recommendation I: Get the word out.* This recommendation calls for the distribution of the recommendation-related information on a broad scale, aimed at raising effective awareness across the oil and gas industrial sector.
- *Recommendation II: Apply appropriate technologies.* This recommendation calls for actions such as follows:
  - Evaluate with care technological solutions for the targeted areas.
  - Use the most suitable fall protection equipment and technology available to the upstream industrial sector.
  - Make use of fire and high-pressure protection systems for controlling struck-by and fire hazards.
  - Apply currently available technology for preventing dropped objects.
  - Use real-time data-transmission technologies for reducing the number and frequency of personnel at well sites.
  - Implement the use of flame-retardant protective clothing.

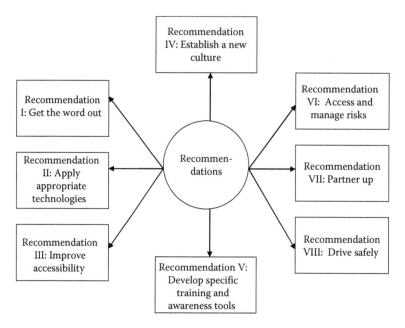

FIGURE 7.4
Recommendations to reduce fatal oil and gas industry incidents.

- *Recommendation III: Improve accessibility.* This recommendation calls for increasing the availability of education tools, training, and information to all involved. Also, incorporating evidence of cost reduction and increment in productivity into these tools.

- *Recommendation IV: Establish a new safety culture.* This recommendation calls for all involved companies (i.e., both contractors and users) for establishing minimum expectations for safety performance by including elements such as follows:
  - Personnel protective equipment
  - Training
  - Equipment standards

  It is to be noted that the effective use of leading indicators can be quite helpful to improve the health, safety, and environment culture.

  Examples include attendance at safety meetings, conducting pre-job safety briefings, conducting and reviewing safety observations, and completing required safety training.

- *Recommendation V: Develop specific training and awareness tools.* This recommendation calls for, in the industry sector where such tools do not exist, industry personnel to work cooperatively for developing tools targeting the causes that lead to the most fatalities (i.e., struck by objects, falls from heights, and explosions and fires).

- *Recommendations VI: Access and manage risks.* This recommendation calls for appropriately identifying and managing risks through job safety analyses, hazards identification, and so on. More specifically, it calls for actions such as follows:
  - Continuously make workers and others aware that oil and gas E&P sites contain many hazards.
  - Make all workers and visitors aware of the increased risks around rigs.
  - Make standard operation practice for reporting all incidents, subsequent investigations to determine their root causes, and appropriate corrective actions.
- *Recommendation VII: Partner up.* This recommendation calls for joining forces across the industry with OSHA and other interested parties for developing, promoting, and distributing fatality reduction-related awareness and training tools.
- *Recommendation VII: Drive safely.* This recommendation calls for the need for using sound journey management programs, including defensive driver training, elimination of distractions (e.g., eating or use of cell phones or other electronic devices), fatigue management, and use of seat belts.

PROBLEMS

1. Write an essay on human factors contribution to accidents in the oil and gas industry.
2. Discuss the following factors in regard to human factors that affect safety in the nuclear industry:
   - Organizational factors
   - Group factors
3. What were the factors that, directly or indirectly, played an important role in the occurrence of the Piper Alpha disaster?
4. What were the categories in the German nuclear power plants' study that divided human factors that affect safety?
5. What were the categories under which the British Airways' Human Factors Reporting Program "BASIS" divided accident causation?
6. What are the main categories of the causes of accidents in the offshore oil industry? Discuss all of these categories in more detail.
7. What were the main causes of the 254 fatalities the U.S. exploration and production industry experienced onshore?

8. Discuss the following recommendations to reduce fatal oil and gas industry incidents:
   - Apply appropriate technologies
   - Improve accessibility
9. Describe the following terms used in this chapter:
   - Job factors
   - Personnel factors
10. Discuss the following two recommendations for reducing fatal oil and gas industry incidents:
    - Develop specific training and awareness tools
    - Access and manage risks

References

1. Cullen, W.D., *The Public Inquiry into the Piper Alpha Disaster, Vols. I and II*, Her Majesty's Stationery Office (HMSO), London, 1980.
2. An Analysis of Root Causes in 1983 Significant Event Reports, Report No. INPO 85-027, Institute of Nuclear Power Operations (INPO), Atlanta, Georgia, 1985.
3. Gordon, R.P.E., The contribution of human factors to accidents in the offshore oil industry, *Reliability Engineering and System Safety*, 61, 1998, 95–108.
4. Donald, I., Canter, D., Employee attitudes and safety in the chemical industry, *Journal of Loss Prevention in the Process Industries*, 17(3), 1994, 50–57.
5. Zohar, D., Safety climate in industrial organizations: Theoretical and applied implications, *Journal of Applied Psychology*, 65(1), 1980, 96–102.
6. Pate-Cornell, M.E., Risk analysis and risk management for offshore platforms: Lessons from the Piper Alpha accident, *Journal of Offshore Mechanics and Arctic Engineering*, 115, 1993, 179–190.
7. Pate-Cornell, M.E., Learning from the Piper Alpha accident: A post-mortem analysis of technical and organizational factors, *Risk Analysis*, 13(2), 1993, 215–232.
8. Kletz, T.A., On the need to publish more case histories, *Plant/Operations Progress*, 7(3), 1988, 145–147.
9. Reason, J., *Human Error*, Cambridge University Press, Cambridge, UK, 1991.
10. Taggart, W., Crew resource management in the cockpit, *Air Line Pilot*, 53, 1984, 20–23.
11. Miller, R., Freitag, M., Wilpert, B., Development and test of a classification scheme for human factors in incident reports, *Proceedings of the IAEA Technical Committee Meeting on Organizational Factors Influencing Human Performance in Nuclear Power Plants*, July 1995, pp. 10–15.
12. Denney, D., Strategic direction for reducing fatal oil and gas industry incidents, *Journal of Petroleum Technology*, 55, July 2005, pp. 66–68.

8

Case Studies of Maintenance Influence on Major Accidents in Oil and Gas Industry and Safety Instrumented Systems and Their Spurious Activation in Oil and Gas Industry

8.1 Introduction

Past experiences indicate that industries such as petroleum and chemical that handle hazardous substances are more prone to major accidents. Maintenance can play a major role in the occurrence of such accidents. Some examples of the major accidents in the area of oil and gas industrial sector in which maintenance, directly or indirectly, has played a major role are Piper Alpha Disaster, Texas City Refinery Explosion, Sodegaura Refinery Disaster, and the Bhopal Gas Tragedy [1].

A safety instrumented system (SIS) may simply be described as a system composed of an engineered set of hardware and software controls that are especially used on critical process systems. More clearly, an SIS is designed to carry out "specific control functions" to fail-safe or maintaining safe operation of a process in the event when unacceptable or dangerous conditions take place. In the oil and gas industrial sector, SISs are used quite frequently on critical process systems.

This chapter presents various important aspects of case studies of maintenance influence on major accidents in the oil and gas industry and of SISs and their spurious activation in the oil and gas industry.

8.2 Maintenance Influence on Major Accidents in the Oil and Gas Industry: Case Studies

Over the years, maintenance, directly or indirectly, has played an important role in the occurrence of major accidents in the oil and gas industry around the globe. The case studies of some of these accidents are presented below, separately [1].

8.2.1 Piper Alpha Accident

The Piper Alpha offshore oil production platform was located about 120 miles northeast of Aberdeen, United Kingdom and operated by Occidental Petroleum (Caledonia) Ltd. The platform became operational in 1976, and it was initially constructed for producing crude oil, but later it also started producing gas with the installation of a gas conversion equipment. At the time of the occurrence of the accident, this platform produced about 10% of North Sea oil and gas [2–4].

On July 6, 1988, the Piper Alpha Offshore platform experienced a series of explosions causing the structural collapse of the platform resulting in 167 deaths [5,6]. This accident is considered to be the worst offshore industry disaster in terms of both lives lost and impact to the offshore industry. In regard to maintenance, some of the factors that directly or indirectly played an important role in the occurrence of this accident were as follows [1]:

- Deficient planning
- Deficient scheduling
- Deficient checking
- Poor maintenance and safety procedures

Additional information on this accident is available in References 1 through 6.

8.2.2 Sodegaura Refinery Accident

This accident occurred on October 16, 1992 when the Sodegaura refinery in Japan, following the breaking-off of the lock ring of the channel cover of heat exchanger and the blowing-off of the lock ring, channel cover, and other parts, experienced an explosion and fire [1,7]. As the result of this accident, 10 people were killed and 11 injured [1,7].

In regard to maintenance, some of the factors that directly or indirectly played an important role in the occurrence of this accident were as follows [1,7]:

- Maintenance error: wrong replacement of the gasket retainer that directly or indirectly contributed to hydrogen gas leakage
- Removal of insulation, which induced temperature difference that resulted in the tube area inner parts' thermal deformation and contributed to the increment of the channel barrel diameter
- Maintenance error: repeated ratcheting, resulting in reduction in the gasket retainer diameter that keeps the heat exchanger air-tight

- Maintenance error: improper internal flange set bolts replacement, leading to their destruction, increment in load on the channel cover set bolts, bending, and diameter reduction in the lock ring

Additional information on Sodegaura Refinery accident is available in References 1 and 7.

8.2.3 Texas City Refinery Accident

This accident occurred on March 23, 2005 at the British Petroleum (BP) Texas City Refinery, the third largest oil refinery in the United States, when a hydrocarbon vapor cloud exploded at the isomerization process unit [8]. More specifically, on March 23, 2005 at the refinery, an isomerization unit's start-up whose raffinate tower was overfilled resulted in the raffinate overheating and pressure relief devices' opening, and then consequently led to a flammable liquid geyser from a blow down stack unequipped with flare and then an explosion and fire [1,9]. The accident killed 15 workers and injured over 170 others [1,8,9].

Some of the active influencing maintenance factors were as follows [1,9]:

- Lack of maintenance: failure to clean sight glass.
- High-level alarm failure, directly or indirectly, due to deficient maintenance program.
- Maintenance error: calibrated level transmitter incorrectly.

Additional information on the Texas City Refinery accident influencing maintenance-related factors is available in References 1, 8, and 9.

8.2.4 Flixborough Accident

This accident occurred on June 1, 1974 at the Flixborough Works of Nypro (United Kingdom) Limited when a bypass system ruptured and released cyclohexane, which in turn formed a combustible mixture with air and then exploded on coming in contact with an ignition source [1,10,11]. As the result of this accident, 28 workers were killed and 36 injured [1,10,11]. Two of the active influencing maintenance factors were as follows [1,10]:

1. Lack of maintenance: after plant modification, the bypass line was not pressure tested.
2. Lack of maintenance: limited calculations were performed on the bypass line.

All in all, the influence of maintenance also included poor planning, poor checking, and poor execution [1]. Additional information on Flixborough accident is available in References 1, 10, and 11.

8.2.5 Bhopal Gas Accident

This accident occurred in the early hours of December 3, 1984 at the Union Carbide pesticide plant in Bhopal, India. The accident released at least 30 tons of a highly toxic gas called methylisocyanate (MIC) as well as a number of other poisonous gases that resulted in about 4000 deaths and 500,000 injuries [1].

In regard to maintenance, some of the active influencing maintenance factors that directly or indirectly played an important role in the occurrence of this accident were as follows [1,5]:

- Nitrogen-line valves' failure due to lack of maintenance
- Product-line valves' failure due to corrosion because of lack of maintenance
- Execution of maintenance initiating a hazardous reaction between MIC and water
- Maintenance error: omission of an isolating blank/spade between the MIC tank and the connected product line under cleaning process using water

Three latent failures were as follows [1,5]:

1. Unbalanced business objectives and cost with maintenance.
2. Unavailable safety features (e.g., the scrubber that should have absorbed the vapor was inoperative and the refrigeration system which could have provided cooling for the storage tank was switched off).
3. Excessive storage of MIC.

All in all, in regard to this accident, the influence of maintenance also included factors such as deficient planning, deficient fault diagnosis, and deficient execution.

8.3 Safety Instrumented Systems

In the oil and gas industry, SISs are used for detecting hazardous events and for performing necessary safety actions for maintaining or bringing the process back to a safe state. An SIS is composed of an engineered set of hardware and software controls, and all the control elements in an SIS are dedicated solely for its proper functioning. These elements (including sensors, actuators, logic solvers, and other control equipment) are of the same types.

An SIS's correct operation requires a series of equipment to operate properly. The specific control functions carried out by a safety instrument system are known as safety instrumented functions (SIF). The overall objective of design, implementation, and follow-up of an SIS is ensuring that the SIS is able to carry out effectively its intended safety functions, when specific process demands occur.

Issues such as those listed below are faced by the SISs market [12]:

- Integration of control and safety
- Standards for programming
- Diagnostic requirements
- Supporting domestic and international standards
- Application fit and function
- Cyber security commitment
- SIS for high availability control, automated startups/shutdowns
- Support and training

Additional information on SISs is available in Reference 13.

8.4 Spurious Activation of Safety Instrumented Systems

Spurious activation of the SISs may result in full or partial shutdown of an ongoing process. In the published literature, spurious activation is known under many different names. Some examples of such names are spurious trip, spurious initiation, premature closure, spurious operation (SO), false trip, spurious activation, and spurious stop [14–17]. In the oil and gas industrial sector, it is very important to reduce or eliminate the occurrence of spurious activations because of the following factors [14]:

- To avoid hazards during unscheduled system restoration and restart
- To avoid unnecessary losses in production
- To reduce the risk associated with stresses caused by the spurious activation

There are three main types of spurious activations as shown in Figure 8.1 [14]. These are SO, spurious trip, and spurious shutdown. An SO is an activation of an SIS element or component without the presence of a stated process demand. Three examples of the SO are as follows [14]:

1. A premature closure of a hydraulically operated, spring loaded, failsafe close safety valve because of leakage in the hydraulic circuit.

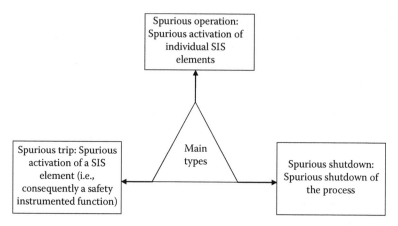

FIGURE 8.1
Three main types of spurious activations.

2. A high-level alarm from a level transmitter without the level of liquid having exceeded its upper limit because of the failure to distinguish the foam from the real level of the liquid in the separator.

3. A false signal about high level from a level transmitter because of an internal malfunction of the transmitter.

A spurious trip is activation of one or more SIS elements or components such that the SIS performs an SIF without the existence of a stated process demand. Two examples of the spurious trip are presented in Reference 14. Finally, a spurious shutdown is a complete or partial process shutdown without the existence of a stated process demand.

There are many causes for the occurrence of SO, spurious trip, and spurious shutdown. Causes for each of these three items are presented below, separately.

8.4.1 Spurious Operation Causes

There are two main causes for SO of a SIS part as shown in Figure 8.2. Additional information on these two main causes is available in Reference 14.

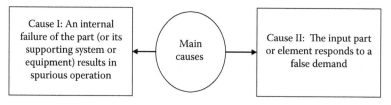

FIGURE 8.2
Main causes of spurious operation of an SIS part or element.

8.4.2 Spurious Trip Causes

SO of SIS parts or elements is one of the main contributors for the occurrence of spurious trips. It may result in a spurious trip if the number of activated parts or elements corresponds to the number of parts or elements required for performing the safety function. Thus, the chosen hardware configuration determines whether or not an SO results in a spurious trip.

It is to be noted that there are several other causes for the occurrence of spurious trips. Some examples are as follows [14]:

- Loss of utilities like power supply, hydraulic or pneumatic. More clearly, this may result in spurious trip when the SIF is designed fail-safe, which is generally the case in the oil and gas industry installations.
- Dangerous detected failures.
- SIF trip due to a human error during the function testing process.

8.4.3 Spurious Shutdown Causes

There are many direct or indirect causes for the occurrence of spurious shutdowns. The main ones are as follows [14]:

- Spurious trips
- Process equipment failures
- Human errors

It is to be noted that a spurious trip generally, but not always, results in spurious shutdown of the ongoing process. More clearly, in regard to process equipment failures, a spurious closure/stop of non-SIS equipment, like control valves and pumps that interact with the ongoing process, may lead to spurious shutdown. The spurious closure of control valves or the spurious stop of pumps may be due to factors such as element internal failures, automatic control system errors, and human errors.

8.5 International Electrotechnical Commission and Its Standards: IEC 61511 and IEC 61508

The International Electrotechnical Commission (IEC) comprising all national electrotechnical committees (i.e., IEC National Committees) is a worldwide organization concerned with standardization. IEC was established in 1906 and its headquarters are located in Geneva, Switzerland. Its objective is to promote effectively international cooperation on all types of questions concerning standardization in the electronic and electrical areas. Additional information on IEC is available in Reference 18.

Two standards (i.e., IEC 61511 and IEC 61508) produced by IEC, directly or indirectly, in regard to SISs are discussed below, separately.

8.5.1 IEC 65111: Functional Safety—Safety Instrumented Systems for the Process Industry Sector

This standard sets out practices in the engineering of systems that ensure an industrial process's safety through the application of instrumentation. The scope of IEC 61511 includes the following items [19,20]:

- Initial concept
- Design
- Implementation
- Operation
- Maintenance through to decommissioning

The standard is composed of three main parts as shown in Figure 8.3 [19].

For an identified SIS, IEC 61511 requires a management system. The SIS management system should define with care how an operator or owner intends to assess, design, engineer, verify, install, commission, validate, operate, maintain, and on a continuous basis improve its SIS. The essential roles of the personnel responsible for the SIS should be outlined with care and procedure developed, as appropriate, for supporting the consistent execution of their assigned responsibilities.

Additional information on IEC 61511 is available in References 19 and 20.

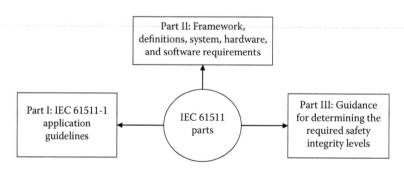

FIGURE 8.3
IEC 61511 standard main parts.

8.5.2 IEC 61508: Functional Safety of Electrical/Electronic/ Programmable Electronic Safety-Related Systems

This standard is intended to be a basic functional safety standard that is applicable to all types of industries. Functional safety in IEC 61508 is defined as follows [21]:

> Part of the overall safety relating to the equipment under control (EUC) and the EUC control system which depends on the correct functioning of the electrical/electronic/programmable electronic safety-related systems, other technology safety-related systems and external risk reduction facilities.

IEC 61508 sets out the requirements to ensure that systems are designed, implemented, operated, and maintained for providing the required system integrity level (SIL). Four SIL are defined according to the level of risks involved in regard to the system application, with SIL4 being used against the highest level of risks [22]. The standard clearly outlines a procedure that can be followed through by all links involved in the supply chain, so that necessary information concerning the system can be communicated using common terminology and parameters associated with the system.

IEC 61508 is composed of eight parts as presented in Table 8.1 [22].

The standard has the following views concerning risks [21]:

- Safety must be considered right from the initial stage.
- All nontolerable risks must be minimized/reduced.
- It is not possible to reach zero risk.

Finally, it is added that some of the sector-specific standards related to IEC 61508 are as follows [22]:

TABLE 8.1

IEC 61508 Parts

Part No.	Part Description
1	IECTR 61508-0, Functional safety and IEC 61508
2	IEC 61508-1, General requirements
3	IEC 61508-2, Requirements for E/E/PE safety-related systems
4	IEC 61508-3, Software requirements
5	IEC 61508-4, Definitions and abbreviations
6	IEC 61508-5, Examples and methods for the determination of safety integrity levels
7	IEC 61508-6, Guidelines on the application of IEC 61508-2 and IEC 61508-3
8	IEC 61508-7, Overview of techniques and measures

- IEC 61513: Nuclear power plants
- IEC 61511: Process industries
- IEC 62061: Machinery sector

Additional information on IEC 61508 is available in References 22 and 23.

PROBLEMS

1. Describe safety instrumented systems.
2. Discuss the maintenance influence on the following two accidents:
   - Piper Alpha accident
   - Sodegaura Refinery accident
3. What are the main issues faced by the safety instrumented systems market?
4. What are the main types of spurious activation of safety instrumented systems?
5. Discuss the maintenance influence on the following two accidents:
   - Texas City Refinery accident
   - Bhopal Gas accident
6. What are the causes for the spurious trip of the safety instrumented systems?
7. Write an essay on International Electrotechnical Commission (IEC).
8. Describe IEC 65111 standard.
9. Write an essay on IEC 61508 standard.

References

1. Okoh, P., Haugen, S., The influence of maintenance on selected major accidents, *Chemical Engineering Transactions*, 31, 2013, 493–498.
2. Petrie, J.R., Piper Alpha Technical Investigation Interim Report, Petroleum Engineering Division, Department of Energy, London, UK, 1988.
3. Pate-Cornell, M.E., Learning from the Piper Alpha accident: Analysis of technical and organizational factors, *Risk Analysis*, 13(2), 1993, 215–232.
4. Dhillon, B.S., *Mine Safety: A Modern Approach*, Springer-Verlag, London, 2010.
5. Kletz, T., *Learning from Accidents*, Gulf Professional Publishing, Oxford, UK, 2001.
6. Pate-Cornell, M.E., Risk analysis and risk management for offshore platforms: Lessons from the Piper Alpha accident, *Journal of Offshore Mechanics and Artic Engineering*, 115(1), 1993, 179–190.

7. Failure Knowledge Database (FKD), Case details, retrieved on June 7, 2012 from website: http://www.sozogaku.com/fkd/en/cfen/CB1011018.html.

8. Texas City Refinery Explosion, retrieved on March 23, 2015 from website: https://en.wikipedia.org/wiki/Texas_City_Refinery_explosion (last modified on December 31, 2015).

9. Texas City Refinery Explosion, Report No. 2005-04-1-TX, United States Chemical Safety Board, Washington, DC, 2007.

10. European Agency for Safety and Health at Work (EU-OSHA): Flixborough Accident, retrieved on June 29, 2012 from website: http://www.hse.gov.uk/comah/sragtech/caseflixboroug74.htm.

11. Flixborough Case History in Building Process Safety Culture: Tools to Enhance Process Safety Performance, Center for Chemical Process Safety of the American Institute of Chemical Engineers, New York, 2005.

12. Safety Instrument System (SIS) Evaluation and Selection, retrieved on April 5, 2015 from website: http://www.arcweb.com/technology-evaluation-and-selection/pages/safety-instrumented-systems.

13. Gruhn, P., Cheddie, H., *Safety Instrumented Systems: Design, Analysis, and Justification*, The Instrumentation, Systems, and Automation Society, Research Triangle Park, North Carolina, 2006.

14. Lundteigen, M.A., Rausand, M., Spurious activation of safety instrumented systems in the oil and gas industry: Basic concepts and formulas, *Reliability Engineering and System Safety*, 93, 2008, 1208–1217.

15. Petroleum, Petrochemical, and Natural Gas Industries-Collection and Exchange of Reliability and Maintenance Data for Equipment, ISO 14224, International Standards Organization (ISO), Geneva, Switzerland, 2006.

16. Nuclear Power Plants-Instrumentation and Control Functions Important for Safety-Use of Probabilistic Safety Assessment for the Classification, IEC TR 61838, International Electrotechnical Commission (IEC), Geneva, Switzerland, 2001.

17. Nuclear Power Plants-Instrumentation and Control for Systems Important to Safety-General Requirements for Systems, IEC 61513, International Electrotechnical Commission, Geneva, Switzerland, 2001.

18. International Electro technical Commission (IEC), retrieved on April 30, 2015, from website: https://en.wikipedia.org/wiki/International_Electrotechnical_Commission (last modified on November 25, 2015).

19. IEC 61511: Functional Safety-Safety Instrumented Systems for the Process Industry Sector, retrieved on January 26, 2015 from website: https://en.wikipedia.org/wiki/IEC_61511 (last modified on August 19, 2015).

20. IEC 61511-1: Safety Instrumented Systems for the Process Industry Sector, International Electrotechnical Commission, Geneva, Switzerland, 2003.

21. IEC 61508: Functional Safety of Electrical/Electronic/Programmable Electronic Safety-Related Systems, retrieved on January 26, 2015 from website: https://en.wikipedia.org/wiki/IEC_61508 (last modified on December 11, 2015).

22. What is IEC 61508?, retrieved on May 1, 2015 from website: http://www.61508.org/?page-id=18.

23. Catelani, M., Ciani, L., Luongo, V., Safety Analysis in Oil and Gas Industry in Compliance with Standards IEC 61508 and IEC 61511: Methods and Application, *Proceedings of the IEEE International Instrumentation and Measurement Technology Conference*, 2013, pp. 686–690.

9

Oil and Gas Industry Accident Data and Accident Data Analysis

9.1 Introduction

The oil and gas industry is a major industrial sector in which a large number of accidents occur each year around the globe. Companies in the oil and gas industry have been collecting accident data for many years for purposes such as monitoring their safety performance and in some countries, to comply with government requirements.

In 1985, the international association of exploration and production oil and gas companies started to collect international accident-related data from its member companies [1]. Initially, 22 companies submitted their data, and the number has grown over to 49 companies located around the globe [2]. The definitions of terms and the type of data to be collected were agreed upon in the committee meetings. The members of these committees represented a wide spectrum of companies from various countries [1].

This chapter presents various important aspects of the oil and gas industry accident data and accident data analysis.

9.2 Offshore Oil and Gas Industry Accident Databases and Accident Data Collection Sources

There are many databases and sources that can be used to obtain offshore oil and gas industry accident-related information.

The databases are generally developed to meet legislative requirements and usually collect data in the continental shelf of their respective country (e.g., countries in the European Union).

Nonetheless, some of the databases and sources considered useful for obtaining offshore oil and gas industry accident-related information are described below, separately [3].

9.2.1 Worldwide Offshore Accident Databank

Worldwide Offshore Accident Databank (WOAD) is one of the main sources for obtaining offshore accident-related information for public use. It is operated by Det Norske Veritas (DNV) of Norway. Most of the data contained in the WOAD is from the Norwegian and UK sectors and the U.S. Gulf of Mexico, and it contains more than 6000 events from 1975 onward, including near misses, incidents, and accidents [3].

In this databank, data are derived mainly from public domain sources, including official publications, newspapers, and Lloyds Casualty Reports (UK). WOAD contains data on items such as follows [3]:

- Type and operation mode of the unit involved in the accident
- Geographical location
- Date
- Main event and chain of events
- Causes and consequences
- Evacuation details

It is to be noted that in this database, exposure data are also provided that allow accident rates to be calculated for different accident and installation/platform/rig types. Finally, it is added that this databank data are not publicly available, but are accessible through a database subscription with a certain amount of charges.

Additional information on WOAD is available at the following website [3]: http://www.dnvusa.com/Binaries/flyer-WOAD tcm153-136061.pdf.

9.2.2 Well Control Incident Database

The starting point for this database was the formation of the Wells Committee of the International Association of Oil & Gas Producers (IOGP). The main objective of this committee was to highlight areas for improvement and then focus on these areas for strengthening the long-term health of the oil and gas industrial sector across the entire cycle of well planning, construction, operation, and abandonment. Some of the objectives of this committee were as follows [3]:

- Provide a formal and active body through which its members can share good practices for contributing to IOGP objectives concerning well integrity-related matters and its mission to facilitate continuous improvements in the area of safety and the environments
- Analyze all incidents and disseminate lessons learned and good practices based on shared experience among its all members

Thus, to meet objectives such as these, this database has been developed. All members of IOGP report their well control incidents and near misses into this database. All data submitted are anonymous and confidential, and the data are only available to members, but not to the public.

Additional information on this database is available at the following website [3]: http://www.ogp.org.uk/committees/wells.

9.2.3 Collision Database

This database was created in 1985 in the United Kingdom for the Health and Safety Executive (HSE), Offshore Safety Division (OSD) [3]. The database was originally concerned with vessel/platform collision incidents on the United Kingdom Continental Shelf (UKCS). Over the years, it has been amended and extended several times.

The database includes those incidents that have been defined as a reported impact between a mobile or fixed installation and a vessel. Although, accident reports or accident data are not available to the public, but the HSE provides from this database quite comprehensive reports with accident statistics and accident frequencies.

Additional information on this database is available at the following website [3]: http://www.hse.gov.uk/research/rrhtm/rr053.htm.

9.2.4 Hydrocarbon Release Database

This database was developed in the United Kingdom, and it contains supplementary information dating back to October 1992, on all types of offshore hydrocarbons releases reported to the HSE OSD under the Reporting of Injuries, Diseases and Dangerous Occurrences Regulations (RIDDOR) 1995 and earlier offshore legislation. The information is submitted voluntarily on the form called OIR/12 form and is also recorded in a separate and specifically designed database maintained by the HSE OSD.

Only authorized users can log on to this database. Additional information on hydrocarbon release database is available at the following website [3]: http://www.hse.gov.uk/hcr3/index.asp.

9.2.5 Danish Energy Agency

Danish Energy Agency (DEA) is a very good source for obtaining offshore oil and gas industry accident-related data. In Denmark, accidents and near misses are reported to DEA using a special notification form or the electronic reporting system called EASY. As per government regulations, the employer or company in charge of operating the offshore installation must register the following items [3]:

- All near-miss incidents occurring on the offshore installation
- All accidents or fatalities occurring on the offshore installation

- Any significant damage to the offshore installation structure or equipment

In addition, the company or employer liable in regard to providing protection (i.e., the company or employer in whose business or service the accident occurred) must report the following information to DEA as per the regulations [3]:

- Any accident that results in incapacity to work for more than one day beyond the injury date
- Accidents that result in fatalities

Each year, the DEA compiles statistics on all reportable accidents and near-miss incidents and publishes them in the annual report on oil and gas production in Denmark. The DEA makes use of these statistics and the individual reports concerning injuries and near-miss incidents to prioritize its supervision-related activities.

Finally, it is to be noted that all the accident reports are not directly available to the public. Additional information on DEA is available at the following website [3]: http://www.ens.dk/en-US/OilAndGas/Health-and-Safety/Work-Related-injuries%20etc/Sider/Forside.aspx.

9.2.6 Performance Measurement Project

Performance Measurement Project (PMP) is another very good source for obtaining offshore oil and gas industry accident-related data. The PMP was set up by the International Regulators' Forum (IRF) for measuring and comparing offshore safety performance among IRF participants by collecting and comparing incident-related data on the basis of a common set criteria.

Data include items such as collisions, fatalities, injuries, fires, gas releases, and well control losses.

Names of the IRF members and their respective countries that provided the data are presented in Table 9.1 [3].

9.2.7 International Association of Oil & Gas Producers (IOGP)

This association is a very good source for obtaining offshore and onshore oil and gas industry accident-related data. The association was formed in 1974, and today most of the world's leading publicly traded, private and state-owned oil and gas companies, oil and gas associations, and major upstream companies are its members [4]. These members produce more than half of the oil of the world and around one-third of its gas.

Since 1985, IOGP has been collecting safety incident-related data from its member companies, and in 1987, it published a first comprehensive collection on global safety performance indicators [4–6]. IOGP performance indicators

TABLE 9.1

Names of the IRF Members and Their Respective Countries That Provided the Data

No.	IRF Member: Name	Country
1	US Bureau of Safety and Environmental Enforcement (BSEE)	United States
2	Newfoundland Labrador Offshore Petroleum Board (NLOPB)	Canada
3	Nova Scotia Offshore Petroleum Board (NSOPB)	Canada
4	Health and Safety Executive (HSE)	United Kingdom
5	National Offshore Petroleum Safety and Environmental Management Authority (NOPSEMA)	Australia
6	New Zealand Department of Labour (DOL)	New Zealand
7	Danish Energy Authority (DEA)	Denmark
8	Petroleum Safety Authority (PSA)	Norway
9	State Supervision of Mines (SSM)	The Netherlands
10	National Petroleum Agency (ANP)	Brazil
11	National Hydrocarbons Commission (CNH)	Mexico

include items such as number of fatalities, fatal accident rate, fatal incident rate, total recordable injury rate, and lost time injury frequency [5].

Additional information on IOGP in regard to accident-related data is available at the following website [5]: http://www.iogp.org/data-series.

9.3 Onshore and Offshore Oil and Gas Industry Accident Data and Analysis

Onshore and offshore oil and gas industry safety incident data have been collected and entered into its safety database by IOGP from its member companies since 1985. Today, this is the largest database of safety performance in the exploration and production industrial sector [2]. Thus, this section presents the data and its analysis from this database for 2012 [2].

Accident-related data and its analysis are presented below under three different categories [2].

9.3.1 Category I: Fatalities

In 2012, there were 88 fatalities (80 onshore and 8 offshore) that occurred in 52 separate incidents. The activity with the highest fatalities was "maintenance, inspection, and testing," in which 41 fatalities occurred in nine separate incidents. These incidents included a gas leak and explosion following the loss of mechanical integrity of a pipeline in Mexico that resulted in 31 fatalities. The activity "construction, commissioning, and decommissioning" accounted for 14 fatalities.

The percentage breakdown for the 88 fatalities in regard to activities was as follows [2]:

- Maintenance, inspection, and testing: 47.7%
- Construction, commissioning, and decommissioning: 16.3%
- Drilling, work over, and well services: 12.8%
- Transport—load: 10.5%
- Production operations: 3.5%
- Seismic/survey operations: 2.3%
- Transport—air: 2.3%
- Transport—water, including marine activity: 2.3%
- Unspecified—other: 2.3%

9.3.2 Category II: Lost Time Injuries

In 2012, there were 1699 reported injuries that resulted in at least one day off work, and a total of 53,325 lost work days were reported. The percentage breakdowns for the lost work day cases in regard to activities were as follows [2]:

- Drilling, work over, and services: 21.2%
- Maintenance, inspection, and testing: 16.9%
- Production operations: 12.4%
- Unspecified—other: 12.4%
- Construction, commissioning, decommissioning: 9.9%
- Lifting, crane, rigging, and deck operations: 7.8%
- Office, warehouse, accommodation, and catering: 7.7%
- Transport—water, including marine activity: 5.2%
- Transport—land: 4%
- Seismic/survey operations: 1.4%
- Subsea diving: 0.6%
- Transport—air: 0.5%

9.3.3 Category III: Fatal Accident Rate

In 2012, the fatal accident rate per 100 million hours worked for onshore and offshore was as follows [2]:

- Onshore: 2.87
- Offshore: 0.89

It is to be noted that over a 10-year period (i.e., up to 2012), there was a large variation between the onshore and offshore fatal accident rate [2]. Furthermore, neither was consistently lower. Generally, this was attributable to single transportation or fire and explosion incidents resulting in a high number of fatalities.

9.4 Offshore Oil and Gas Rigs Accident Analysis

Offshore oil and gas industry accident-related data have been collected and entered into the WOAD since 1970. It is considered one of the most complete and reliable databases of accidents, incidents, and failures in the offshore oil and gas industry. Therefore, this section presents offshore oil and gas rigs accident analysis of data obtained from WOAD.

As per Reference 3, WOAD contains a total of 6183 records concerning near misses, incidents, accidents, etc. Geographical distribution of these records is as follows [3]:

- North Sea: 3505 records
- Gulf of Mexico: 1685 records
- Mediterranean Sea: 45 records
- All other regions of the world (i.e., South America, Africa, and Australia): 866 records

The distribution of the 45 Mediterranean Sea records is shown in Figure 9.1 [3].

The records contained in the WOAD database are classified under the following four categories [3]:

- *Category I: Accidents.* They represent hazardous situations that have developed into accidental conditions. All events/situations resulting in fatalities and severe injuries are covered under this category.
- *Category II: Incidents/hazardous situations.* They represent hazardous situations that did not develop into accidental conditions. More clearly, in such situations, low degree of damage was recorded, but generally replacements/repairs were required. Also, the events causing minor injuries to personnel or health injuries are included under this category.
- *Category III: Near misses.* They represent events that may have developed into accidental situations. More clearly, in these cases, there was no damage and no repairs were required.

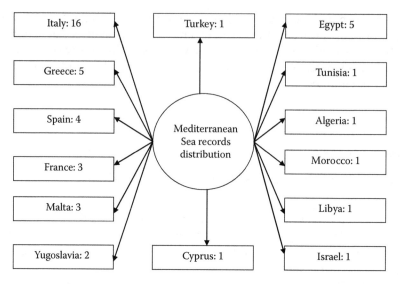

FIGURE 9.1
Distribution of Mediterranean Sea records concerning near misses, incidents, and accidents.

- *Category IV: Insignificant events.* They represent hazardous conditions/ situations with very low consequences. More clearly, in most of the cases, no damages were registered and repairs were not carried out. This category also includes small spills of crude oil and chemicals, in addition to very minor personnel injuries (i.e., "lost time incidents").

The percentage of 6183 records of the above four categories of events are shown in Figure 9.2 [3].

Distribution of accidents per type of unit for accidents in the WOAD database is as follows [3]:

- Fixed units: 50%
- Mobile units: 38%
- Other: 12%

9.4.1 Distribution of Accidents per Type of Human-Related Causes

Over the years, many accidents have occurred in the offshore oil and gas industrial sector due to human-related causes. As per WOAD database, the distribution of accidents per type of human-related causes is as follows [3]:

- Unsafe act/no procedure: 44%
- Unsafe procedure: 37%
- Third-party error: 9%

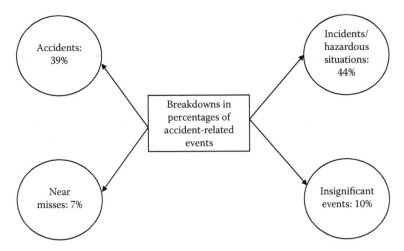

FIGURE 9.2
Percentages of 6183 records of four types/categories of accident-related events.

- Improper design: 8%
- Act of war/during warsit/sabotage: 1%
- Other: 1%

As per the above distribution of causes, it is to be noted that the causes that mostly contribute to the occurrence of events are associated with procedures either as unsafe procedures (i.e., 37% of events) or as absence of procedures that lead to unsafe acts (i.e., 44% of events).

Finally, it is added that as per the WOAD database, there is a very large percentage of events (i.e., 5323 events [approximately 86% of the cases]) in which no human-related causes were attributed [3].

9.4.2 Distribution of Accidents per Type of Equipment-Related Causes

Over the years, there have been many accidents in the offshore oil and gas industry due to equipment-related causes. As per the WOAD database, the distribution of accidents per type of equipment-related causes is as follows [3]:

- All equipment malfunction (electrical and mechanical): 34%
- Ignition (all types of ignition included [i.e., open flame, weld, heat, hand tools and sparks, torch, electrical, lightning, cigarette/match, and unknown]): 26%
- Weather, general: 25%
- Foundation and structure failure (including corrosion and fatigue): 8%
- Third-party equipment malfunction: 5%

- Safety system failure: 0.18%
- Earthquake, volcanic eruption: 0.18%
- Other (including exceeding design criteria): 2%

The above distribution of percentages shows that 85% of per type equipment-related causes are all equipment malfunction (34%), ignition (26%), and weather, general (25%). The causes related to safety systems' failure are very rare (i.e., 0.18%).

Finally, it is to be noted that as per the WOAD database, in approximately 55% of the cases (i.e., 3355 events out of a total of 6183 recorded events), no equipment-related causes were attributed to the events [3].

9.5 Failures and Lessons Learned from Landmark Offshore Oil and Gas Accidents and Corrective Measures

The study of offshore oil and gas accidents indicates that there have been various types of failures that directly or indirectly led to the occurrence of these accidents. The lessons learned for controlling the relevant risk for the occurrence of such accidents and keeping them to an adequately low level could be very useful to the operators, the regulators, and the international community for their necessary corrective measures.

In this section, the failures that led to such accidents are grouped under eight classifications along with their corresponding corrective measures that will help to eliminate or reduce such failures [3]. These eight classifications are presented below, separately.

9.5.1 Classification I: Prevention

Three types of failures that belong to this classification are shown in Figure 9.3 [3].

Type I failures can be prevented with the performance of adequate risk assessment through actions such as follows [3]:

- For hazard identification, existence, application, and review of good-quality standards
- Identification of hazards during all phases of the oil and gas exploitation activity life cycle, during changes of procedures and boundary conditions, and under extreme conditions

Similarly, Type II failures can be prevented with appropriate cementing of the well through actions such as follows [3]:

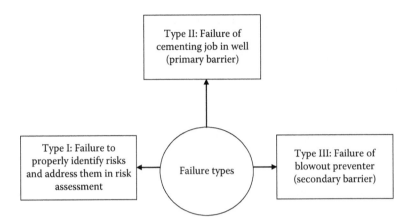

FIGURE 9.3
Classification I failures types.

- Proper oversight by all involved regulatory authorities; review risk assessment, check capacity of operator/contractor, and control conformity
- Operator (i.e., organization/company) follows proper procedures
- Existence of high-level well integrity-related practices and standards
- Operator maintains high level of safety culture
- Contractor/operator recognizes early signals and reacts instantly

Finally, Type III failures can be prevented with the installation of a blowout preventer (BOP) with adequate features that ensure performance as preventer and integrate in prevention system, with actions such as follows [3]:

- Operator applies recognized practices and the state-of-the-art technology
- Regulatory authority regularly performs inspections, oversees risks, and reviews risk management
- Existence of high level technology-related standards

9.5.2 Classification II: Mitigation

Three types of failures that belong to this classification are as follows [3]:

- Type I: Failure to protect vulnerable areas such as control room, vulnerable compartments, and workers' area from the explosion impact

- Type II: Failure to properly use the diverter; overreliance on human response under pressure
- Type III: Failure to avoid ignition of released hydrocarbons

Type I failures can be prevented with the use of materials and designs that can withstand increased overpressure through actions such as follows [3]:

- Operator installs appropriate state-of-the-art protection measures
- Existence of good practices and best technologies to protect vulnerable areas
- Regulatory authority regularly checks protection measures' adequacy

Similarly, Type II failures can be prevented with the installation of a diverter of appropriate design and with the adequate features and ensuring that in case of accident, it is used in the proper way, with actions such as follows [3]:

- Operator uses recognized best practices and state-of-the-art technology
- Existence of good-quality technology-related standards with proper balance between human/automatic intervention
- Regulatory authority regularly performs inspections, oversees risks, and reviews risk management
- Risk assessment ensures appropriate protection level and better reliability of the overall protection system

Type III failures can be prevented with the installation and functioning of gas detectors in properly defined hazardous areas, to avoid ignition sources in these areas, through actions such as follows [3]:

- Operator installs appropriate state-of-the-art gas detectors in appropriate locations and extends the areas considered hazardous as the need arises
- Regulatory authority performs inspections and checks protection measures' adequacy
- Existence of rather good practices for the definition of hazardous and high technology in gas detectors

9.5.3 Classification III: Emergency

The failure that belongs to this classification is "failure to properly respond to the accident" [3]. This failure can be prevented with the application of highly sophisticated emergency response technologies, application of efficient plans, and mobilizing all required capacities of the operators and the member organizations/countries, with actions such as follows [3]:

- Existence of high-level technology-related standards and best available technologies for responding to emergencies
- Emergency plan with the involvement of various authorities from affected member organizations/countries
- Existence of adequate capacities

9.5.4 Classification IV: Safety Management

The failure that belongs to this classification is "failure to manage safety of operations properly" [3]. The occurrence of this failure can be prevented by putting in place a safety and environmental management system and addressing systematically and continuously the operations' safety challenges, with actions such as follows [3]:

- Operator uses/applies best-recognized practices
- Regulatory authority monitors the level of safety and reviews safety management systems as necessary
- Existence of good safety-related practices
- Operator takes appropriate actions to enhance and promote safety culture within its organization

9.5.5 Classification V: Preparedness and Planning

The failure that belongs to this classification is "failure to be properly prepared to respond to the accident" [3]. This failure can be prevented by being prepared and foreseeing the capacities required to respond to the accident and developing a plan on how to respond, through actions such as follows [3]:

- Operator develops emergency plan (internal) on the basis of good practices and commonly acceptable scenarios
- Development of scenarios and assessment of capacities appropriate for responding to these scenarios (e.g., to contain the spill, to stop the release, and to rescue personnel)
- Regulatory authority must review and inspect all the emergency plans and confirm the appropriate existence of capacities
- Existence of good-quality practices

9.5.6 Classification VI: Aftermath/Restoration

The failure that belongs to this classification is "failure to restore the environment to the status prior to the accident" [3]. The occurrence of this failure can be prevented by taking appropriate measures to restore the quality of the environment, with actions such as follows [3]:

- Operator applies appropriate recognized best practices and state-of-the-art technology
- Regulatory authority monitors and oversees all types of cleanup operations
- Existence of good-quality technology standards for cleanup-related operations

9.5.7 Classification VII: Early Warning

The failure that belongs to this classification is "failure to recognize and react to early warning signals of hydrocarbons entering the well" [3]. The occurrence of this failure can be prevented by better monitoring, early detection, and interpretation of early warning signals, through the following action [3]:

- Application and existence of recognized good practices.

9.5.8 Classification VIII: Lesson Learning

The failure that belongs to this classification is "failure to learn from near misses and from accidents" [3]. The occurrence of this failure can be prevented by putting in place an appropriately designed system for investigating accidents, identifying key lessons, and learning lessons from near misses, incidents, and accidents, with actions such as follows [3]:

- Operator investigates all near misses, incidents, and accidents, and then identifies lessons learned and disseminates them to personnel within the company/organization, other operators, risk management community, and inspectors
- Proper existence of a common format to report near misses, incidents, and accidents
- Regulatory authority collects data regularly and then forwards it to the commission for performing further analysis
- Commission (or other independent body) performs analysis of accidents and then disseminates the lessons learned
- Proper existence of all agreed taxonomies of the causes, consequences, and critical issues concerned with them, as well as lessons learned

PROBLEMS

1. Describe the Worldwide Offshore Accident Databank (WOAD).

2. What is the International Regulators' Form (IRF) and who are its members?

3. Write an essay on the International Association of Oil & Gas Producers (IOGP).

4. What are the classifications of failures that directly or indirectly lead to the occurrence of accidents in the offshore oil and gas industrial sector?

5. What is the distribution of Mediterranean Sea records concerning near misses, incidents, and accidents in regard to offshore oil and gas industry?

6. How many records are contained in WOAD and what is their geographical distribution?

7. Write an essay on the Danish Energy Agency (DEA).

8. Describe the Well Control Incident Database.

9. Compare WOAD and the Well Control Incident Database.

10. Describe the following two databases:

   - Collision Database
   - Hydrocarbon Release Database

References

1. Cloughley, T.M.G., Thomas, I., Accident data for the upstream oil and gas industry, *Proceedings of the 3rd International Conference on Health, Safety, and Environment in Oil and Gas Exploration and Production*, 1996, pp. 161–169.

2. Safety Performance Indicators—2012 Data, Report No. 2012s, International Association of Oil & Gas Producers, London, UK, June 2012.

3. Christon, M., Konstantinidou, M., Safety of Offshore Oil and Gas Operations: Lessons from Past Accident Analysis, Joint Research Center Scientific and Policy Report No. 77767, European Commission, Luxembourg, 2012.

4. International Association of Oil and Gas Producers, retrieved on May 11, 2015 from website: https://en.wikipedia.org/wiki/International_Association_of_Oil_%26_Gas_Producers (last modified on December 2, 2014).

5. International Association of Oil and Gas Producers, IOGP Data Series, retrieved on May 11, 2015, from website: http://www.iogp.org/data-series.

6. International Association of Oil and Gas Producers, About IOGP, retrieved on May 11, 2015 from website: http://www.iogp.org/About-IOGP.

10

Oil and Gas Industry Equipment Reliability

10.1 Introduction

Billions of dollars are spent globally every year to construct/manufacture, operate, and maintain various types of equipment/systems for the oil and gas industry. Reliability of equipment used in the oil and gas industrial sector has become an important issue due to various types of equipment reliability-related problems over the years. For example, corrosion-related failures including maintenance, direct cost in the U.S. petroleum industry alone was $3.7 billion per year in 1996 [1,2].

Today, the global economy is forcing the companies involved with oil and gas to modernize their operations through increased automation and mechanization. Thus, as equipment used in the oil and gas industry is becoming more sophisticated and complex, its cost is increasing quite rapidly. To meet production targets, oil and gas companies are increasingly demanding better equipment reliability.

This chapter presents various important aspects of oil and gas industry equipment reliability.

10.2 Mechanical Seals' Failures

Mechanical seals have been increasingly used to seal rotating shafts for more than seven decades. Currently, they are the most common types of seals found on items such as centrifugal pumps and compressors used in the oil and gas industry. Over the years, the failure of mechanical seals has become an important issue. For example, a study conducted in a petroleum company reported that 60% of plant breakdowns, directly or indirectly, were due to mechanical seal failures [3].

A study of mechanical seal failures in a company reported the following causes for the failure of mechanical seals [3]:

- Wrong seal spring compression
- Shaft and seal face plane misaligned
- Metal particles embedded in the carbon
- External system or component failure (e.g., bearings)
- Hang-up (i.e., coking)
- Highly worn carbon (i.e., greater than 4 mm of wear)
- Auxiliary seal system failure (e.g., cooling, flush, quench, recirculating)
- Hang-up (i.e., crystallization)
- Dry-running
- Seal component failure (i.e., other than the faces or secondary seals)

10.2.1 Mechanical Seals' Typical Failure Modes and Their Causes

Mechanical seals can fail in many different failure modes due to various causes. Typical failure modes and their corresponding causes for mechanical seals are as follows [4]:

- *Failure mode I: Fractured spring.* Its causes are misalignment, corrosion, material flaws, and stress concentration due to tooling marks.
- *Failure mode II: Seal fracture.* Its causes are excessive fluid pressure on seal, stress-corrosion cracking, and excessive pressure velocity (PV) value.
- *Failure mode III: Accelerated seal face wear.* Its causes are excessive torque, misalignment, inadequate lubrication, shaft out-of-roundness, contaminants, excessive shaft end play, and surface finish deterioration.
- *Failure mode IV: Excessive friction resulting in slow mechanical response.* Its causes are excessive seal swell, metal-to-metal contact (i.e., out of alignment), seal extrusion, and excessive squeeze.
- *Failure mode V: Seal face edge chipping.* Its causes are excessive shaft whip, excessive shaft deflection, and seal faces out-of-square.
- *Failure mode VI: Small leakage.* Its causes are installation damage and insufficient squeeze.
- *Failure mode VII: Compression set and low pressure leakage.* Its cause is extreme temperature operation.
- *Failure mode VIII: Seal face distortion.* Its causes are insufficient seal lubrication, excessive pressure on seal, excessive PV value of seal operation, and foreign material trapped between faces.
- *Failure mode IX: O-ring failure.* Its causes are excessive fluid pressure, installation error, and excessive temperature (i.e., greater than 55°C).

- *Failure mode X: Clogged spring.* Its cause is fluid contaminants.
- *Failure mode XI: Open seal face-axial.* Its causes are temperature growth, spiral failure (caused by conditions which allow some parts of the ring to slide and others to roll that cause twisting), impeller adjustment error, thrust movement, etc.
- *Failure mode XII: Axial shear.* Its cause is excessive pressure loading.
- *Failure mode XIII: Seal embrittlement.* Its causes are idle periods between use, contaminants, thermal degradation, and fluid/seal incompatibility.
- *Failure mode XIV: Fluid seepage.* Its causes are foreign material on rubbing surface and insufficient seal squeeze (loss of spring tension).
- *Failure mode XV: Open seal face-radial.* Its causes are bent shaft, shaft detection, and shaft whip.
- *Failure mode XVI: Torsional shear.* Its causes are excessive fluid pressure surges and excessive torque due to improper lubrication.
- *Failure mode XVII: Clogged bellows.* Its causes are particles stuck at the inside of the bellows and hardening of fluid during downtime.

Additional information on mechanical seal failure modes is available in Reference 4.

10.3 Optical Connector Failures

Optical fiber connectors are used for joining optical fibers in situations where a disconnect/connect capability is required. Fiber-optic equipment including wet-mate optical connectors is an important part of the current subsea infrastructure in oil and gas applications. A study of the reliability data of the optical connectors (excluding cables or jumpers) collected over the 10-year period reported four factors/issues (including their percentage breakdowns), as shown in Figure 10.1, that caused optical connector failures [5]. The data include field failures and the failures that occurred during integration into equipment and testing process before field deployment.

It is to be noted from Figure 10.1 that 86% of the optical connector failures were due to mechanical, material, and external factors, and only 14% of failures were related to optical issues. Furthermore, when only field failures were studied, the optical connector failures occurred due to the following two types of issues only [5]:

1. *Mechanical issues*: They accounted for 61% of the failures.
2. *Material issues*: They accounted for 39% of the failures.

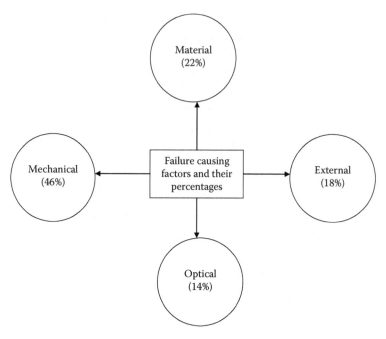

FIGURE 10.1
Factors along with their percentages in parentheses that caused optical connector failures.

   Additional information on optical connector failures is available in Reference 5.

10.4 Corrosion-Related Failures

Corrosion in the oil and gas industry has been acknowledged since the 1920s, and corrosion and other related failures still cost the offshore oil and gas industrial sector hundreds of millions of dollars each year [6]. As per Reference 7, corrosion-related failures constitute over 25% of failures in the oil and gas industrial sector [7]. A study conducted in the 1980s reported the following causes (along with their degree of contribution in percentages) for corrosion-related failure in petroleum-related industries [7,8]:

- Erosion corrosion: 9%
- Stress corrosion: 3%
- CO_2 related: 28%
- Pitting: 12%
- H_2S related: 18%

- Galvanic: 6%
- Impingement: 3%
- Preferential weld: 18%
- Crevice: 3%

10.4.1 Types of Corrosion or Degradation That Can Cause Failure

The possible types of corrosion or degradation that can cause failure are as follows [6]:

- *Corrosion fatigue*: Over the years, it has played an important role in subsurface and drilling operations such as drill pipe and sucker-rod failures.
- *Stress corrosion cracking*: The most probable form of cracking phenomenon in oil and gas production is sulfide and chloride stress corrosion cracking.
- *Erosion–corrosion*: It is often observed on the outer radius of pipe bends in oil and gas production due to rather high fluid flow rates as well as corrosive environments where flow exceeds 6 m/s for copper nickel and 10 m/s for carbon steel.
- *Crevice corrosion*: This type of corrosion occurs in situations where crevice forms, such as partial penetration welds and backing strips, are employed.
- *Weight loss corrosion*: This type of corrosion occurs most commonly in oil and gas production due to an electrochemical reaction between metal and the corrodents in the environment.
- *Fretting corrosion*: It usually occurs in poorly lubricated valve stems where a partially opened valve causes some vibration that can result in galling and then, in turn, valve seizure and possible failure.
- *Hydrogen-induced cracking*: In the past, this type of cracking has mostly occurred in the controlled rolled pipeline steels with elongated stringers of nonmetallic imperfections.
- *Microbiological-induced corrosion*: It is quite serious as it takes the form of localized pitting attack that can cause a rapid loss of metal in a concentrated area, leading to leak or rupture.
- *Galvanic corrosion*: It can occur in bimetallic connections at opposite ends of the galvanic series that have enough potential for causing a corrosion reaction in the existence of an electrolyte.
- *Impingement/cavitation*: Impingement can happen in situations where process fluid is forced to change its flow direction abruptly. Common offshore area for the occurrence of cavitation is in pump impellors where pressure changes take place and high liquid flow rates occur.

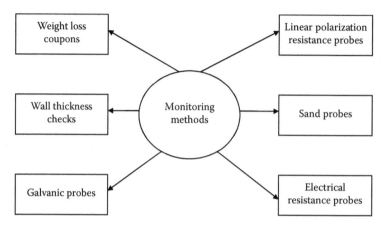

FIGURE 10.2
Commonly used corrosion/condition monitoring methods.

10.4.2 Corrosion/Condition Monitoring Methods

Corrosion monitoring of internal surfaces may be performed by using the combination of the following methods [6]:

- Measurements of nonintrusive wall thickness (radiography/ultrasonic)
- Visual inspection
- Intrusive probes and coupons for monitoring corrosion and erosion
- Pipeline leak detection
- Chemical analysis of samples taken from the product

The commonly used monitoring methods are shown in Figure 10.2 [6].

Additional information on methods shown in Figure 10.2 is available in Reference 6.

10.5 Oil and Gas Pipeline Fault Tree Analysis

The fault tree method was developed at the Bell Telephone Laboratories in the early 1960s to perform an analysis of the Minuteman Launch Control System with respect to safety [9]. Currently, the method is widely used around the globe for performing various types of reliability and safety studies. The method is described in detail in Chapter 4.

Here, the application of this method to perform oil–gas long pipeline failure analysis is demonstrated through the following two examples [10].

EXAMPLE 10.1

Assume that an oil–gas pipeline failure can occur due to any of the following events: pipeline with defects, material with poor mechanical properties, misoperation, pipeline with serious corrosion, and third-party damage. The occurrences of four of these events are described below:

1. The event "pipeline with defects" can be either due to pipeline with construction defects or pipeline with initial defects.
2. The event "misoperation" can be due to design misoperation, operate misoperation, or maintain misoperation.
3. The event "pipeline with serious corrosion" can occur due to the occurrence of the following: corrosion and pipeline with poor corrosiveness resistance. In turn, the event "corrosion" can be either due to external corrosion or internal corrosion.
4. The event "third-party damage" can be due to either artificial damage or natural disaster and external force.

With the aid of fault tree symbols given in Chapter 4, develop a fault tree for the top event "oil–gas pipeline failure."

A fault tree for the example is shown in Figure 10.3. The single capital letters in Figure 10.3 denote corresponding fault events (e.g., X: pipeline with defects, A: material with poor mechanical properties, and Z: misoperation).

EXAMPLE 10.2

Assume that in Figure 10.3, the probabilities of occurrences of fault events A, B, C, D, E, F, G, H, I, J, and K are 0.02, 0.04, 0.05, 0.15, 0.06, 0.08, 0.12, 0.07, 0.14, 0.01, and 0.09, respectively. With the aid of Chapter 4, calculate the probability of occurrence of the top event T: oil–gas pipeline failure.

With the aid of material presented in Chapter 4, we calculate the probabilities of occurrence of fault events N, X, Y, Z, M, and T as follows:

The probability of the occurrence of event N is

$$P(N) = 1 - (1 - P(J))(1 - P(K))$$

$$= 1 - (1 - 0.01)(1 - 0.09) \qquad (10.1)$$

$$= 0.0991$$

where
$P(J)$ is the occurrence probability of event J.
$P(K)$ is the occurrence probability of event K.

Similarly, the probability of the occurrence of event X is

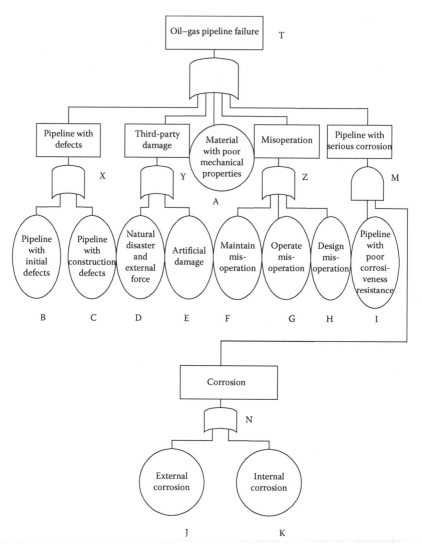

FIGURE 10.3
A fault tree for the top event: Oil–gas pipeline failure.

$$P(X) = 1 - (1 - P(B))(1 - P(C))$$

$$= 1 - (1 - 0.04)(1 - 0.05) \qquad (10.2)$$

$$= 0.088$$

where
  $P(B)$ is the occurrence probability of event B.
  $P(C)$ is the occurrence probability of event C.

Similarly, the probability of the occurrence of event Y is

$$P(Y) = 1 - (1 - P(D))(1 - P(E))$$

$$= 1 - (1 - 0.15)(1 - 0.06) \qquad (10.3)$$

$$= 0.201$$

where
$P(D)$ is the occurrence probability of event D.
$P(C)$ is the occurrence probability of event C.

Finally, the probability of the occurrence of event Z is

$$P(Z) = 1 - (1 - P(F))(1 - P(G))(1 - P(H))$$

$$= 1 - (1 - 0.08)(1 - 0.12)(1 - 0.07) \qquad (10.4)$$

$$= 0.2470$$

where
$P(F)$ is the occurrence probability of event F.
$P(G)$ is the occurrence probability of event G.
$P(H)$ is the occurrence probability of event H.

The probability of the occurrence of the intermediate event M is

$$P(M) = P(N)P(I)$$

$$= (0.0991)(0.14) \qquad (10.5)$$

$$= 0.0138$$

where
$P(I)$ is the occurrence probability of event I.

The top event T (oil–gas pipeline failure) probability of occurrence is

$$P(T) = 1 - (1 - P(X))(1 - P(Y))(1 - P(A))(1 - P(Z))(1 - P(M))$$

$$= 1 - (1 - 0.088)(1 - 0.201)(1 - 0.02)(1 - 0.2470)(1 - 0.0138) \qquad (10.6)$$

$$= 0.4697$$

where
$P(A)$ is the occurrence probability of event A.

Thus, the probability of the occurrence of the top event T (oil–gas pipeline failure) is 0.4697. Figure 10.3 fault tree with given and calculated probability values is shown in Figure 10.4.

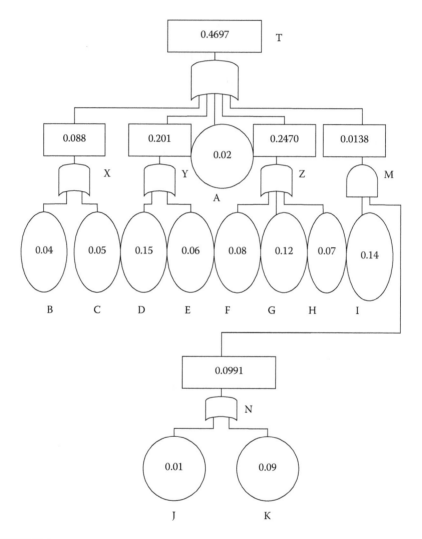

FIGURE 10.4

Redrawn Figure 10.3 fault tree with given and calculated event occurrence probability values.

10.6 Common Cause Failures Defense Approach for Oil and Gas Industry Safety Instrumented Systems

Safety instrumented systems (SIS) in the oil and gas industry generally function in the low demand mode, which means that regular inspection and testing are necessary to reveal their failures. Past experiences indicate that the occurrences of common cause failures are a serious threat to SIS reliability and may result in simultaneous failures of redundant parts/units and safety barriers [11–13].

Thus, a common cause failure may simply be defined as any instance where multiple parts/units/components fail due to a single cause [13]. Some of the causes for the occurrence of common cause failures are design deficiency, operation and maintenance errors, external normal environment, external catastrophe, common manufacturer, and common external power source.

The common cause failures defense approach described below for oil and gas industry SIS focuses on the following three key aspects [11]:

1. To avoid introducing common cause failures during inspection and function testing processes
2. To use the insight of failure causes for choosing efficient mechanisms to defend against future occurrences of common cause failures
3. To highlight common cause failures and their causes on the basis of failure reports

10.6.1 Common Cause Failures Defense Approach

This approach is based on six function testing and inspection tasks: (i) scheduling; (ii) preparation, execution, and restoration; (iii) failure reporting; (iv) failure analysis; (v) implementation; and (vi) validation and continuous improvements. Thus, the approach is composed of six tasks based on checklists and analytical methods, such as influence diagrams, cause-defense matrices, and operational sequence diagrams (OSD). The six tasks are described as follows [11]:

- *Task 1:* This task is concerned with ensuring that all appropriate improvements are captured during the scheduling process. Here, it is to be noted that an important defense against the occurrence of common cause failures is to ensure that any improvements and corrections to the test procedure are clearly captured during the creation of a new function test or inspection work packages.
- *Task 2:* This task is concerned with avoiding the introduction of common cause failures during preparation, execution, and restoration. Three separate checklists, containing questions for preparation, execution and restoration presented below, are considered quite useful [11].

Preparation checklist:

- Are the individuals involved with executing the test clearly familiar with the calibration and testing tools?
- Have all types of potential human errors during execution and restoration been clearly discussed and highlighted?

- Have compensating appropriate measures been clearly highlighted and implemented for avoiding human errors?
- Does the procedure describe essential steps for safely restoring the SIS?
- Are all the calibration tools properly calibrated?
- Does the procedure contain known deficiencies (e.g., ambiguous instructions)?
- Have human error-related incidents been experienced during earlier execution?

Execution checklist:

- Are all the parts properly protected against damage from nearby work-related activities?
- Are all the field SIS parts (constituting the safety function under test) properly labeled?
- Are all the parts operated within the stated operating and environmental conditions?
- Are all the additional parts that are operated during SIS function testing and inspection process properly labeled?
- Are the process connections free from plugging and (if applicable) heat-traced?

Restoration checklist:

- Has the safety function been properly verified prior to start-up?
- Has the physical restoration (e.g., bypasses and isolation valves) been properly verified?
- Are any remaining bypasses, inhibits, or overrides logged, and compensating measures properly highlighted and implemented?
- Have all inhibits' and overrides' suspensions been properly verified and communicated?

- *Task 3*: This task is concerned with improving failure reporting quality. In this regard, the following questions are considered useful [11]:
  - What was the effect of the occurrence of failure on the overall safety function (i.e., degraded, loss of entire function, none at all)?
  - How was the failure observed or uncovered (i.e., during the inspection or repair process, by diagnostics, by review, incidentally, upon demand, or during function testing)?

- Has the part been overexposed (i.e., environmental or by operational stresses); if so, what could be the associated causes?
- What appears to be the failure cause(s)?
- Have similar types of failures occurred previously?
- Was the part inspected or tested in a different way than outlined in the inspection or test procedure; if so, what was the reason for the approach to be different?

- *Task 4*: This task is concerned with identifying common cause failures through failure analysis. The following four steps are considered quite useful in identifying common cause failures [11]:

  - *Step 1*: Review the description of the failure and verify the initial failure classification (if necessary correct it).

  - *Step 2*: Conduct an appropriate initial screening that captures failures that (i) clearly share failure-related causes, (ii) have been found within the framework of the same test or inspection interval, (iii) have very similar physical location or design, and (iv) the causes for failures are not random as defined by IEC 61508, 1998 and IEC 61511m 2003 documents [11].

  - *Step 3*: Conduct a root cause and coupling factor analysis with the aid of influence diagrams.

  - *Step 4*: List all the root cause and coupling factors in a cause-defense matrix.

- *Task 5*: This task is concerned with implementing defensive measures. The proper implementation of common cause failures-related defensive measures is very important for preventing failure occurrences of similar types of failures. Additional information on this task is available in Reference 11.

- *Task 6*: This task is concerned with validation and continuous improvements. In regard to validation, the following questions are considered useful [11]:

  - Are all disciplines concerned with SIS inspection, testing, maintenance, and follow-up properly familiar with the common cause failures concept?

  - Are all changes in operating or environmental conditions properly captured and analyzed for essential modifications to the SIS or associated procedures?

  - Are common cause failures systematically highlighted and analyzed, and appropriate defenses implemented for preventing their reoccurrences?

  - Are all requirements for the safety function properly covered by the inspection or function test procedure(s)?

- Are individuals using the test and calibration tools clearly familiar with their proper application?
- Are all the test-related limitations (compared to the actual demand conditions) clearly known?
- Are all the procedure-related shortcomings properly communicated to the responsible individuals and followed up?
- Are all failures introduced during function testing and inspection processes captured, analyzed, and used for improving the related procedures?
- Are all dangerous undetected failure modes clearly known and appropriately catered for in the function test and inspection-related procedures?
- Are all the diagnostic alarms properly followed up within the started mean time to restoration?
- Are all failures detected upon real demands properly analyzed for verifying that they would have been detected during an inspection or function test?
- Are all safety function redundant channels properly covered by the function test or inspection-related procedures?
- Are all the calibration and test tools suitable and maintained as per the vendor recommendations?

Finally, it is added that for the above questions, the answer "no" indicates a potential weakness in the defense against the occurrence of common cause failures, and should be discussed for determining appropriate corrective measures.

10.7 Fatigue Damage Initiation in Oil and Gas Steel Pipes Assessment

Currently, steel pipes are commonly used in offshore petroleum industry. In regard to the dynamic loadings (i.e., cyclic loadings), the fatigue behavior is a very important issue of concern. As per References 14 through 16, fatigue is one of the main failure causes observed in oil and gas steel pipes, which in turn can result in catastrophic environmental damage and significant financial losses.

In order to assure structural integrity of oil and gas steel pipes and forewarn a fatigue-related failure, it is very important to adopt a consistent fatigue criterion. Nonetheless, the fatigue-related damage may be divided into the following two main phases [14]:

1. *Incubation phase*: During this phase, only microstructural changes, microcracks nucleation, and microcracking can be observed. It is to be noted that the study of this phase is more cumbersome to perform since microstructural changes and fatigue damage cannot be easily separated.

2. *Propagation phase*: During this phase, macroscopic cracking and macrocrack propagation result in fatigue failure, and the physical data that may quantify the material's damage state can be more easily obtained.

Nondestructive evaluation (NDE) methods are very useful for evaluating structural integrity and assessing fatigue life. These methods can assess the limitation of fatigue damage, monitor changes in mechanical properties, and follow the fatigue damage process through the structures' life cycle subjected to cyclic loadings, in order to forewarn a failure. Some of these methods are as follows [14]:

- X-ray diffraction method
- Ultrasonics method
- Magnetic evaluation method
- Thermography method
- Hardness measurements method

It is to be noted that among the nondestructive methods for fatigue damage imitation assessments, x-ray diffraction method is considered to be one of the most suitable analysis methods [14]. Additional information on this method is available in Reference 14.

PROBLEMS

1. What are the causes for the failure of mechanical seals? List at least eight such causes.
2. List at least 10 typical failure modes of mechanical seals.
3. Discuss optical connector failures.
4. Discuss at least 10 types of corrosion or degradation that can cause failure.
5. What are the commonly used corrosion/condition monitoring methods?
6. Assume that in Figure 10.3, the probabilities of occurrences of fault events A, B, C, D, E, F, G, H, I, J, and K are 0.03, 0.05, 0.06, 0.16, 0.07, 0.09, 0.13, 0.08, 0.16, 0.08, and 0.07, respectively. Calculate the probability of occurrence of the top event: oil–gas pipeline failure and also the oil–gas pipeline reliability.

7. What are the causes for the occurrence of common cause failures?

8. Describe the common cause failures defense approach.

9. Discuss fatigue damage initiation in oil and gas steel pipes assessment?

10. What are the causes for the occurrence of following failure modes of mechanical seals:

   - Seal fracture
   - Excessive friction
   - O-ring failure

References

1. Oil Refinery, retrieved on January 13, 2015 from website: https://en.wikipedia.org/wiki/Oil_refinery (last modified on January 19, 2016).

2. Kane, R.D., Corrosion in petroleum refining and petrochemical operations, in *Metals Handbook, Vol. 13C: Environments and Industries*, edited by S.O. Cramer and B.S. Covino, ASM International, Metals Park, Ohio, 2003, pp. 967–1014.

3. Wilson, B., Mechanical seals, *Industrial Lubrication and Tribology*, 47(2), 1995, 4.

4. Skewis, W.H., Mechanical Seal Failure Modes, Support Systems Technology Corporation, Gaithersburg, Maryland, retrieved on May 28, 2015 from website: http://docslide.us/documents/mechanical-seal-failure-modes.html.

5. Jones, R.T., Thiraviam, A., Reliability of fiber optic connectors, *Proceedings of the IEEE OCEANS Conference*, 2010, pp. 1–10.

6. Price, J.C., Fitness-for-purpose failure and corrosion control management in offshore oil and gas development, *Proceedings of the 11th International Offshore and Polar Engineering Conference*, 2001, pp. 234–241.

7. Kermani, M.B., Harrop, D., The Impact of Corrosion on the Oil and Gas Industry, Society of Petroleum Engineers (SPE) Production and Facilities, August 1996, pp. 186–190.

8. Kermani, M.B., Hydrogen cracking and its mitigation in the petroleum industry, *Proceedings of the Conference on Hydrogen Transport and Cracking in Metals*, 1994, pp. 1–8.

9. Dhillon, B.S., Singh, C., *Engineering Reliability: New Techniques and Applications*, John Wiley and Sons, New York, 1981.

10. Tian, H. et al., Application of fault tree analysis in the reliability analysis of oil-gas long pipeline, *Proceedings of the International Conference on Pipelines and Trenchless Technology*, 2013, pp. 1436–1446.

11. Lundteigen, M.A., Rausand, M., Common cause failures in safety instrumented systems on oil and gas installations: Implementing defense measures through testing, *Journal of Loss Prevention in the Process Industries*, 20, 2007, 218–229.

12. Summers, A.E., Raney, G., Common cause and common sense, designing failure out of your Safety Instrumented System (SIS), *ISA Transactions*, 38, 1999, 291–299.

13. Dhillon, B.S., Proctor, C.L., Common-mode failure analysis of reliability networks, *Proceedings of the Annual Reliability and Maintainability Symposium*, 1977, pp. 404–408.

14. Pinheiro, B. et al., Assessment of fatigue damage initiation in oil and gas steel pipes, *Proceedings of the ASME 30th International Conference on Ocean, Offshore, and Arctic Engineering*, 2011, pp. 1–10.

15. Lyons, D., Western European Cross-Country Oil Pipelines 30-Year Performance Statistics, Report No. 1/02, CONCAWE, Brussels, Belgium, 2002.

16. Paulson, K., *A Comparative Analysis of Pipeline Performance, 2000–2003*, National Energy Board, Alberta, Canada, 2005.

11

Mathematical Models for Performing Safety and Reliability Analyses in Oil and Gas Industry

11.1 Introduction

Mathematical modeling is a widely used method for performing various types of analyses in the area of engineering. In such models, system parts/components are represented by idealized elements that are assumed to have all the representative characteristics of real-life parts/components, and whose behavior is possible to be represented by equations. However, a mathematical model's degree of realism very much depends on the types of assumptions imposed upon it.

Over the years, a large number of mathematical models have been developed for studying the safety, reliability, and maintainability of systems in the area of engineering. Many of these models were developed using stochastic processes, including the Markov method [1–3]. Although the effectiveness of such models can vary from one application area to another, some of these models are being used quite successfully to study various types of real-life problems in the industrial sector [4–6]. Thus, some of these mathematical models can also be used for studying safety and reliability-related problems in the oil and gas industrial sector.

This chapter presents a number of mathematical models considered useful to perform various types of safety and reliability-related analyses in oil and gas industry.

11.2 Model I

This mathematical model represents a worker performing a time-continuous oil and gas industry-related task subjected to noncritical (safe) errors and critical (unsafe) errors. More clearly, the errors committed by the worker in the oil and gas industry are separated into two groups: noncritical (safe) and critical (unsafe). The state space diagram for the worker performing a time-continuous oil and gas industry-related task is shown in Figure 11.1 [7,8].

The numeral and small single letters in the boxes in Figure 11.1 denote the states of the worker. The following assumptions are associated with this model:

- The worker performs a time-continuous task.
- Noncritical and critical error rates of the worker are constant.
- All errors occur independently.

The following symbols are associated with this model:

j is the jth state of worker performing an oil and gas industry-related task, where $j = 0$ means that the worker is performing an oil and gas industry-related task correctly, $j = s$ means that the worker committed a noncritical (safe) error, and $j = u$ means that the worker committed a critical (unsafe) error.

$P_j(t)$ is the probability of the worker being in state j at time t; for $j = 0$, s, u.

λ_s is the worker constant noncritical (safe) error rate.

λ_u is the worker constant critical (unsafe) error rate.

With the aid of the Markov method described in Chapter 4, we write down the following system of differential equations for Figure 11.1:

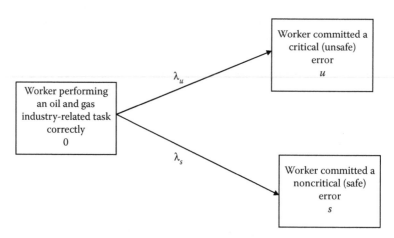

FIGURE 11.1
State space diagram for the worker performing an oil and gas industry-related task.

$$\frac{dP_0(t)}{dt} + (\lambda_s + \lambda_u)P_0(t) = 0 \tag{11.1}$$

$$\frac{dP_s(t)}{dt} - \lambda_s P_0(t) = 0 \tag{11.2}$$

$$\frac{dP_u(t)}{dt} - \lambda_u P_0(t) = 0 \tag{11.3}$$

At time $t = 0$, $P_0(0) = 1$, $P_s(0) = 0$, and $P_u(0) = 0$.
By solving Equations 11.1 through 11.3, we obtain

$$P_0(t) = e^{-(\lambda_s + \lambda_u)t} \tag{11.4}$$

$$P_s(t) = \frac{\lambda_s}{\lambda_s + \lambda_u}[1 - e^{-(\lambda_s + \lambda_u)t}] \tag{11.5}$$

$$P_u(t) = \frac{\lambda_u}{\lambda_s + \lambda_u}[1 - e^{-(\lambda_s + \lambda_u)t}] \tag{11.6}$$

The worker reliability is given by

$$\begin{aligned} R_w(t) &= P_0(t) \\ &= e^{-(\lambda_s + \lambda_u)t} \end{aligned} \tag{11.7}$$

where
$R_w(t)$ is the worker reliability at time t.

The mean time to worker error (MTTWE) is given by [3,7,8]

$$\begin{aligned} \text{MTTWE} &= \int_0^\infty R_w(t)\,dt \\ &= \int_0^\infty e^{-(\lambda_s + \lambda_u)t}\,dt \\ &= \frac{1}{\lambda_s + \lambda_u} \end{aligned} \tag{11.8}$$

EXAMPLE 11.1

Assume that a worker is performing an oil and gas industry-related task and his/her noncritical (safe) and critical (unsafe) error rates are 0.009 and 0.005 errors/h, respectively. Calculate the probability of the worker committing a noncritical (safe) error during an 8-h work period.

By inserting the specified data values into Equation 11.5, we get

$$P(8) = \frac{0.009}{(0.009 + 0.005)} [1 - e^{-(0.009+0.005)(8)}]$$

$$= 0.0681$$

Thus, the probability of the worker committing a noncritical (safe) error during the specified work period is 0.0681.

11.3 Model II

This mathematical model represents a system used in the oil and gas industry having three distinct states: working normally, failed safely, and failed unsafely. The failed system (i.e., safely or unsafely) is repaired back to its normal working state. The oil and gas industry system state space diagram is shown in Figure 11.2 [8,9].

The numeral and small single letters in the circles in Figure 11.2 denote system states. The following five assumptions are associated with this model:

- The oil and gas industry system can fail either safely or unsafely.
- All failures occur independently.
- The oil and gas industry system safe and unsafe failure rates are constant.
- The failed oil and gas industry system repair rates are constant.
- The repaired oil and gas industry system is as good as new.

The following symbols are associated with this model:

i is the ith state of the oil and gas industry system, where $i = 0$ means the system is working normally, $i = s$ means the system failed safely, and $i = u$ means the system failed unsafely.

$P_i(t)$ is the probability that the oil and gas industry system is in state i at time t, for $i = 0, s, u$.

λ_i is the oil and gas industry system's ith constant failure rate, where $i = s$ means safe and $i = u$ means unsafe.

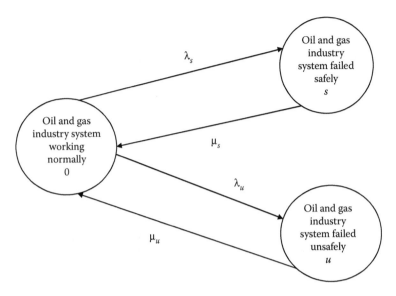

FIGURE 11.2
Oil and gas industry system state space diagram.

μ_i is the failed oil and gas industry system's ith constant repair rate, where $i = s$ means from a safe failed state and $i = u$ means from an unsafe failed state.

With the aid of the Markov method described in Chapter 4, we write down the following system of differential equations for Figure 11.2:

$$\frac{dP_0(t)}{dt} + (\lambda_u + \lambda_s)P_0(t) = \mu_u P_u(t) = \mu_s P_s(t) \tag{11.9}$$

$$\frac{dP_u(t)}{dt} + \mu_u P_u(t) = \lambda_u P_0(t) \tag{11.10}$$

$$\frac{dP_s(t)}{dt} + \mu_s P_s(t) = \lambda_s P_0(t) \tag{11.11}$$

At time $t = 0$, $P_0(0) = 1$ and $P_u(0) = P_s(0) = 0$.
By solving Equations 11.9 through 11.11, we obtain

$$P_0(t) = \frac{\mu_u \mu_s}{X_1 X_2} + \left[\frac{(X_1 + \mu_s)(X_1 + \mu_u)}{X_1(X_1 - X_2)}\right]e^{X_1 t} - \left[\frac{(X_2 + \mu_s)(X_2 + \mu_u)}{X_2(X_1 - X_2)}\right]e^{X_2 t} \tag{11.12}$$

where

$$X_1, X_2 = \frac{-A \pm \sqrt{A^2 - 4(\mu_s\mu_u + \lambda_s\mu_u + \lambda_u\mu_u)}}{2}$$

$$A = \mu_s + \mu_u + \lambda_u + \lambda_s$$

$$X_1 X_2 = (\mu_u\mu_s + \lambda_s\mu_u + \lambda_u\mu_s)$$

$$X_1 + X_2 = -(\mu_s + \mu_u + \lambda_s + \lambda_u)$$

$$P_u(t) = \frac{\lambda_u\mu_s}{X_1 X_2} + \left[\frac{(\lambda_u X_1 + \lambda_u\mu_s)}{X_1(X_1 - X_2)}\right]e^{X_1 t} - \left[\frac{(\mu_s + X_2)\lambda_u}{X_2(X_1 - X_2)}\right]e^{X_2 t} \qquad (11.13)$$

$$P_s(t) = \frac{\lambda_s\mu_u}{X_1 X_2} + \left[\frac{(\lambda_s X_1 + \lambda_s\mu_u)}{X_1(X_1 - X_2)}\right]e^{X_1 t} - \left[\frac{(\mu_u + X_2)\lambda_s}{X_2(X_1 - X_2)}\right]e^{X_2 t} \qquad (11.14)$$

Equations 11.13 and 11.14 give the probability of the oil and gas industry system failing unsafely and safely, respectively, when subjected to the repair process.

As time t becomes large, the oil and gas industry system steady-state probability of failing safely using Equation 11.14 is

$$P_s = \lim_{t\to\infty} P_s(t) = \frac{\lambda_s\mu_u}{X_1 X_2} \qquad (11.15)$$

Similarly, as time t becomes very large, the oil and gas industry system steady-state probability of failing unsafely using Equation 11.13 is

$$P_u = \lim_{t\to\infty} P_u(t) = \frac{\lambda_u\mu_s}{X_1 X_2} \qquad (11.16)$$

By setting $\mu_s = \mu_u = 0$ in Equations 11.9 through 11.11 and then solving the resulting equations, we get

$$P_0(t) = e^{-(\lambda_u + \lambda_s)t} \qquad (11.17)$$

$$P_u(t) = \frac{\lambda_u}{\lambda_u + \lambda_s}[1 - e^{-(\lambda_u + \lambda_s)t}] \qquad (11.18)$$

$$P_s(t) = \frac{\lambda_s}{\lambda_u + \lambda_s}[1 - e^{-(\lambda_u + \lambda_s)t}] \tag{11.19}$$

Equation 11.17 gives the oil and gas industry system reliability at time t. In contrast, Equations 11.18 and 11.19 give the probability of the oil and gas industry system failing unsafely and safely at time t, respectively.

By integrating Equation 11.17 over the time interval $[0,\infty]$, we get the following equation for the oil and gas industry system mean time to failure [7,8]:

$$\begin{aligned} \text{MTTF}_{0s} &= \int_0^\infty P_0(t)\,dt \\ &= \int_0^\infty e^{-(\lambda_u + \lambda_s)t}\,dt \\ &= \frac{1}{\lambda_u + \lambda_s} \end{aligned} \tag{11.20}$$

where
MTTF_{0s} is the oil and gas industry system mean time to failure.

EXAMPLE 11.2

Assume that a system used in the oil and gas industry can fail safely or unsafely and its constant failure rates are 0.0008 and 0.0002 failures/h, respectively. Calculate the probabilities of the system failing safely and unsafely during a 200-h mission.

By substituting the given data values into Equations 11.19 and 11.18, we obtain

$$P_s(200) = \frac{0.0008}{0.0002 + 0.0008}[1 - e^{-(0.0002 + 0.0008)(200)}]$$

$$= 0.1450$$

and

$$P_u(200) = \frac{0.0002}{0.002 + 0.0008}[1 - e^{-(0.0002 + 0.0008)(200)}]$$

$$= 0.0362$$

Thus, the probabilities of the oil and gas industry system failing safely and unsafely during the specified mission period are 0.1450 and 0.0362, respectively.

11.4 Model III

This mathematical model represents a worker performing a time-continuous oil and gas industry-related task under fluctuating environments (i.e., normal and abnormal or stressful) [8,10]. The worker can commit an error in either environment (i.e., normal or abnormal). The state space diagram for the worker performing a time-continuous oil and gas industry-related task under fluctuating environments is shown in Figure 11.3 [8,10].

This model is subjected to the following assumptions:

- All errors and changes in an environment occur independently.
- The worker is performing a time-continuous task.
- The rate of changing the environment from normal to abnormal (or stressful) or vice versa is constant.
- The error rates of the worker are constant.

The numerals in the boxes in Figure 11.3 denote the states of the worker. The following symbols are associated with this model:

i is the ith state of worker performing an oil and gas industry-related task, where $i = 0$ means that the worker is performing an oil and gas industry-related task correctly in a normal environment, $i = 1$ means

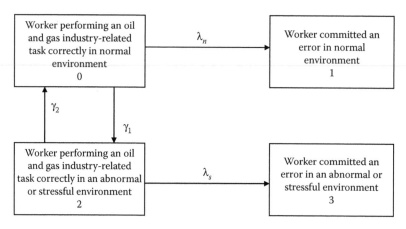

FIGURE 11.3
State space diagram for the worker performing an oil and gas industry-related task.

that the worker committed an error in a normal environment, $i = 2$ means that the worker is performing an oil and gas industry-related task correctly in an abnormal or stressful environment, and $i = 3$ means that the worker committed an error in an abnormal or stressful environment.

$P_i(t)$ is the probability of the worker being in state i at time t, for $i = 0$, 1, 2, 3.

λ_n is the constant error rate of the worker performing an oil and gas industry-related task in a normal environment.

λ_s is the constant error rate of the worker performing an oil and gas industry-related task in an abnormal or stressful environment.

γ_1 is the constant transition rate from a normal environment to an abnormal or stressful environment.

γ_2 is the constant transition rate from an abnormal or stressful environment to a normal environment.

With the aid of the Markov method described in Chapter 4, we write down the following system of differential equations for Figure 11.3:

$$\frac{dP_0(t)}{dt} + (\lambda_n + \gamma_1)P_0(t) = \gamma_2 P_2(t) \tag{11.21}$$

$$\frac{dP_1(t)}{dt} = \lambda_n P_0(t) \tag{11.22}$$

$$\frac{dP_2(t)}{dt} + (\lambda_s + \gamma_2)P_2(t) = \gamma_1 P_0(t) \tag{11.23}$$

$$\frac{dP_3(t)}{dt} = \lambda_s P_2(t) \tag{11.24}$$

At time $t = 0$, $P_0(0) = 1$, $P_1(0) = P_2(0) = P_3(0) = 0$.
By solving Equations 11.21 through 11.24, we get

$$P_0(t) = (y_1 - y_2)^{-1}[(y_2 + \lambda_s + \gamma_2)e^{y_2 t} - (y_1 + \lambda_s + \gamma_2)e^{y_1 t}] \tag{11.25}$$

where

$$y_1 = \frac{-a_1 + \sqrt{a_1^2 - 4a_2}}{2}$$

$$y_2 = \frac{-a_1 - \sqrt{a_1^2 - 4a_2}}{2}$$

$$a_1 = \lambda_n + \lambda_s + \gamma_1 + \gamma_2$$

$$a_2 = \lambda_n(\lambda_s + \gamma_2) + \gamma_1 \lambda_s$$

$$P_1(t) = a_4 + a_5 e^{y_2 t} - a_6 e^{y_1 t} \tag{11.26}$$

where

$$a_3 = \frac{1}{y_2 - y_1}$$

$$a_4 = \frac{\lambda_n(\lambda_s + \gamma_2)}{y_1 y_2}$$

$$a_5 = a_3(\lambda_n + a_4 y_1)$$

$$a_6 = a_3(\lambda_n + a_4 y_2)$$

$$P_2(t) = \gamma_1 a_3 (e^{y_2 t} - e^{y_1 t}) \tag{11.27}$$

$$P_3(t) = a_7[(1 + a_3)(y_1 e^{y_2 t} - y_2 e^{y_1 t})] \tag{11.28}$$

where

$$a_7 = \frac{\lambda_s \gamma_1}{y_1 y_2}$$

The worker reliability in a fluctuating environment is given by

$$R_{wf}(t) = P_0(t) + P_1(t) \tag{11.29}$$

where
$R_{wf}(t)$ is the worker reliability of performing an oil and gas industry-related task in a fluctuating environment at time t.

The mean time to worker error in a fluctuating environment is given by

$$\text{MTTWE}_f = \int_0^\infty R_{wf}(t)\,dt$$

$$= \frac{(\lambda_s + \gamma_1 + \gamma_2)}{a_2}$$

(11.30)

where
   MTTWE_f is the mean time to worker error in a fluctuating environment.

EXAMPLE 11.3

Assume that a worker is performing an oil and gas industry-related task in a fluctuating (i.e., normal and stressful) environment and his/her constant error rates are 0.003 and 0.009 errors/h, respectively. The constant transition rates from normal to stressful environment and from stressful to normal environment are 0.005 and 0.002 per hour, respectively. Calculate his/her mean time to error.

   By inserting the specified data values into Equation 11.30, we get

$$\text{MTTWE}_f = \frac{(0.009 + 0.005 + 0.002)}{0.003(0.009 + 0.002) + (0.005)(0.009)}$$

$$= 205.1\,\text{h}$$

Thus, mean time to his/her error is 205.1 h.

11.5 Model IV

This model is concerned with predicting the reliability of a worker performing an oil and gas industry-related time-continuous task. In this case, the worker reliability is expressed by [3,11,12]

$$R_w(t) = e^{-\int_0^t \alpha_w(x)\,dx}$$

(11.31)

where
   $R_w(t)$ is the oil and gas industry worker reliability at time t.
   $\alpha_w(x)$ is the oil and gas industry worker instantaneous error rate.

   It is to be noted that in Equation 11.31, the time to oil and gas industry worker's human error can follow any time-continuous probability distribution (e.g., exponential, normal, and Weibull).

EXAMPLE 11.4

Assume that an oil and gas industry worker is performing a time-continuous task and his/her error rate is 0.002 errors/h (i.e., time to error of the worker is exponentially distributed). Calculate the reliability of the worker during a 7-h work period.

By inserting the given data values into Equation 11.31, we get

$$R_w(7) = e^{-\int_0^7 (0.002)dx}$$

$$= e^{-(0.002)(7)}$$

$$= 0.9860$$

Thus, the reliability of the worker during the specified work period is 0.9860.

11.6 Model V

This mathematical model represents an equipment used in the oil and gas industry that can fail safely or unsafely. When the equipment fails safely or unsafely, the repair is attempted. If it cannot be properly repaired in the field, it is taken to the repair workshop [2,8,13]. The state space diagram for the oil and gas industry equipment is shown in Figure 11.4.

The numeral and small single letters in the boxes in Figure 11.4 denote the equipment states. The model is subjected to the following assumptions:

- Equipment failure and repair rates are constant.
- The rates of taking the failed equipment to the repair workshop are constant.
- Failures occur independently.
- The repaired equipment is as good as new.

The following symbols are associated with this model:

i is the ith state of the oil and gas industry equipment, where $i = 0$ means that the oil and gas industry equipment is operating normally, $i = u$ means that the oil and gas industry equipment failed unsafely, $i = s$ means that the oil and gas industry equipment failed safely, and $i = w$ means that the oil and gas industry equipment is in the repair workshop.

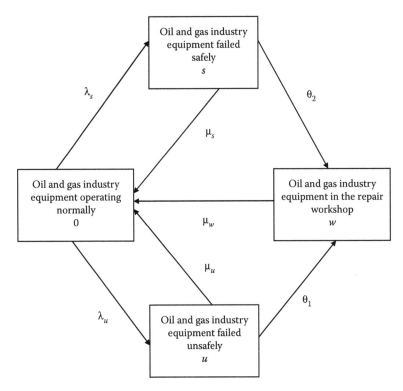

FIGURE 11.4
State space diagram for an oil and gas industry equipment failing safely or unsafely and taken to a repair workshop.

$P_i(t)$ is the probability that the oil and gas industry equipment is in state i at time t, for $i = 0, u, s, w$.

λ_s is the oil and gas industry equipment constant safe failure rate.

λ_u is the oil and gas industry equipment constant unsafe failure rate.

θ_1 is the constant rate of taking the unsafely failed oil and gas industry equipment to the repair workshop.

θ_2 is the constant rate of taking the safely failed oil and gas industry equipment to the repair workshop.

μ_u is the constant repair rate of the oil and gas industry equipment from the unsafely failed state (i.e., state u).

μ_s is the constant repair rate of the oil and gas industry equipment from the safely failed state (i.e., state s).

μ_w is the constant repair rate of the oil and gas industry equipment from the repair workshop (i.e., state w).

With the aid of the Markov method described in Chapter 4, we write down the following system of differential equations for Figure 11.4 [2,8,13]:

$$\frac{dP_0(t)}{dt} + (\lambda_s + \lambda_u)P_0(t) = \mu_u P_u(t) + \mu_s P_s(t) + \mu_w P_w(t) \tag{11.32}$$

$$\frac{dP_u(t)}{dt} + (\theta_1 + \mu_u)P_u(t) = \lambda_u P_0(t) \tag{11.33}$$

$$\frac{dP_s(t)}{dt} + (\theta_2 + \mu_s)P_s(t) = \lambda_s P_0(t) \tag{11.34}$$

$$\frac{dP_w(t)}{dt} + \mu_w P_w(t) = \theta_2 P_s(t) + \theta_1 P_u(t) \tag{11.35}$$

At time $t = 0$, $P_0(0) = 1$, $P_u(0) = P_s(0) = P_w(0) = 0$.

By setting the derivatives equal to zero in Equations 11.32 through 11.35 and using the relationship $P_0 + P_u + P_s + P_w = 1$, we get the following steady-state probability equations [14]:

$$P_0 = \frac{A}{B} \tag{11.36}$$

and

$$P_i = m_i P_0, \text{ for } i = u, s, w \tag{11.37}$$

where

$$A = \mu_w(\theta_1 + \mu_u)(\theta_2 + \mu_s)$$

$$B = (\theta_1 + \mu_u)[\mu_w(\theta_2 + \mu_s) + \lambda_s(\mu_w + \theta_2)] + \lambda_u(\theta_2 + \mu_s)(\theta_1 + \mu_w)$$

$$m_u = \frac{\lambda_u}{(\theta_1 + \mu_u)}$$

$$m_s = \frac{\lambda_s}{(\theta_2 + \mu_s)}$$

$$m_w = \frac{\lambda_u \theta_1 (\theta_2 + \mu_s) + \lambda_s \theta_2 (\theta_1 + \mu_u)}{\mu_w (\theta_1 + \mu_u)(\theta_2 + \mu_s)}$$

P_j is the steady-state probability of the oil and gas industry equipment being in state i, for $i = 0, u, s, w$.

The oil and gas industry equipment steady-state availability and unavailability are given by

$$\text{OGIEA}_s = P_0 \tag{11.38}$$

and

$$\text{OGIEUA}_S = P_u + P_s + P_w \tag{11.39}$$

where

OGIEA_s is the oil and gas industry equipment steady-state availability.

OGIEUA_s is the oil and gas industry equipment steady-state unavailability.

11.7 Model VI

This mathematical model represents a system used in the oil and gas industry that can either fail safely or fail with an accident due to hardware failures or human errors [3,8,15]. When the system fails safely or with an accident, it is taken to the repair workshop for repair. The state space diagram for the oil and gas industry system is shown in Figure 11.5.

The numerals and single letters in the boxes in Figure 11.5 denote system states. The model is subjected to the following assumptions:

- Human error rates, hardware failure rates, and rates of taking the failed oil and gas industry system to the repair workshop are constant.
- Human errors and failures occur independently.
- The repaired oil and gas industry system is as good as new.
- The failed oil and gas industry system repair rate is constant.

The following symbols are associated with the model:

i is the ith state of the oil and gas industry system, where $i = 0$ means that the oil and gas industry system is operating normally, $i = 1$ means that the oil and gas industry system failed safely due to hardware failures, $i = 2$ means that the oil and gas industry failed safely due to human errors, $i = 3$ means that the oil and gas industry system failed with an accident due to hardware failures, $i = 4$ means that the oil and gas industry system failed with an accident due to human errors, and $i = w$ means that the oil and gas industry system equipment is in the repair workshop.

$P_i(t)$ is the probability that the oil and gas industry system is in state i at time t, for $i = 0, 1, 2, 3, 4, w$.

λ_1 is the constant hardware failure rate of the oil and gas industry system failing safely.

λ_2 is the constant hardware failure rate of the oil and gas industry system that causes an accident.

λ_3 is the constant human error rate of the oil and gas industry system failing safely.

λ_4 is the constant human error rate of the oil and gas industry system that causes an accident.

θ_5 is the constant rate of taking the failed oil and gas industry system to the repair workshop for repair from state 1.

θ_6 is the constant rate of taking the failed oil and gas industry system to the repair workshop for repair from state 3.

θ_7 is the constant rate of taking the failed oil and gas industry system to the repair workshop for repair from state 2.

θ_8 is the constant rate of taking the failed oil and gas industry system to the repair workshop for repair from state 4.

μ_w is the constant repair rate of the oil and gas industry system from state w to state 0.

P_i is the steady-state probability that the oil and gas industry system is in state i, for $i = 0, 1, 2, 3, 4, w$.

With the aid of the Markov method described in Chapter 4, we write down the following system of differential equations for Figure 11.5 [3,8,15]:

$$\frac{dP_0(t)}{dt} + (\lambda_1 + \lambda_2 + \lambda_3 + \lambda_4)P_0(t) = \mu_w P_w(t) \tag{11.40}$$

$$\frac{dP_1(t)}{dt} + \theta_5 P_1(t) = \lambda_1 P_1(t) \tag{11.41}$$

$$\frac{dP_2(t)}{dt} + \theta_7 P_2(t) = \lambda_3 P_0(t) \tag{11.42}$$

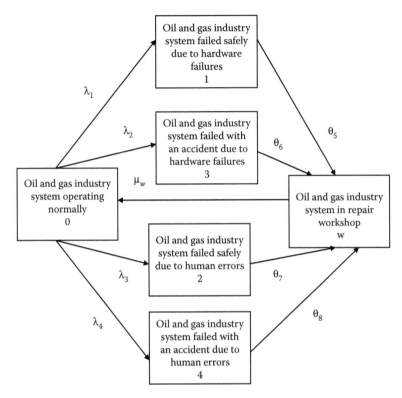

FIGURE 11.5
State space diagram for the oil and gas industry system that can either fail safely or fail with an accident due to human errors or hardware failures.

$$\frac{dP_3(t)}{dt} + \theta_6 P_3(t) = \lambda_2 P_0(t) \tag{11.43}$$

$$\frac{dP_4(t)}{dt} + \theta_8 P_4(t) = \lambda_4 P_0(t) \tag{11.44}$$

$$\frac{dP_w(t)}{dt} + \mu_w P_w(t) = \theta_5 P_1(t) + \theta_6 P_3(t) + \theta_7 P_2(t) + \theta_8 P_4(t) \tag{11.45}$$

At time $t = 0$, $P_0(0) = 1$, $P_1(0) = P_2(0) = P_3(0) = P_4(0) = P_w(0) = 0$.
By setting $\mu_w = 0$ in Equations 11.40 through 11.45 and then solving for $P_0(t)$, we get

$$R_{os}(t) = P_0(t) = e^{-(\lambda_1 + \lambda_2 + \lambda_3 + \lambda_4)t} \tag{11.46}$$

where

$R_{os}(t)$ is the oil and gas industry system reliability at time t.

The oil and gas industry system mean time to failure (MTTF$_{os}$) is given by [14]

$$MTTF_{os} = \int_0^\infty R_{os}(t)\,dt$$

$$= \int_0^\infty e^{-(\lambda_1+\lambda_2+\lambda_3+\lambda_4)t}\,dt \qquad (11.47)$$

$$= \frac{1}{(\lambda_1+\lambda_2+\lambda_3+\lambda_4)}$$

By setting derivatives equal to zero in Equations 11.40 through 11.45 and utilizing the relationship $P_0 + P_1 + P_2 + P_3 + P_4 + P_w = 1$, we obtain the following set of steady-state probability equations [14]:

$$P_0 = \frac{1}{1+N} \qquad (11.48)$$

where

$$N = \frac{\lambda_1}{\theta_5} + \frac{\lambda_3}{\theta_7} + \frac{\lambda_2}{\theta_6} + \frac{\lambda_4}{\theta_8} + \frac{D}{\mu_w}$$

where

$$D = \lambda_1 + \lambda_2 + \lambda_3 + \lambda_4$$

$$P_1 = \frac{\lambda_1}{\theta_5} P_0 \qquad (11.49)$$

$$P_2 = \frac{\lambda_3}{\theta_7} P_0 \qquad (11.50)$$

$$P_3 = \frac{\lambda_2}{\theta_6} P_0 \qquad (11.51)$$

$$P_4 = \frac{\lambda_4}{\theta_8} P_0 \qquad (11.52)$$

$$P_5 = \frac{D}{\mu_w} P_0 \qquad (11.53)$$

The oil and gas industry system steady-state availability and unavailability are given by

$$\text{OGISA} = P_0 \qquad (11.54)$$

and

$$\text{OGISUA} = P_1 + P_2 + P_3 + P_4 + P_w \qquad (11.55)$$

where
OGISA is the oil and gas industry system steady-state availability.
OGISUA is the oil and gas industry system steady-state unavailability.

EXAMPLE 11.5

Assume that in Figure 11.5, for transition rates $\lambda_1, \lambda_2, \lambda_3,$ and λ_4, we have the following specified values:

- $\lambda_1 = 0.0006$ failures/h
- $\lambda_2 = 0.0002$ failures/h
- $\lambda_3 = 0.0005$ failures/h
- $\lambda_4 = 0.0003$ failures/h

Calculate the oil and gas industry system reliability for a 20-h mission and mean time to failure.
By inserting the given data values into Equation 11.46, we obtain

$$R_{os}(20) = e^{-(0.0006+0.0002+0.0005+0.0003)(20)}$$

$$= 0.9685$$

Similarly, by inserting the specified data values into Equation 11.47, we obtain

$$\text{MTTF}_{os} = \frac{1}{(0.0006 + 0.0002 + 0.0005 + 0.0003)}$$

$$= 625\,\text{h}$$

Thus, the oil and gas industry system reliability and mean time to failure for the given data values are 0.9685 and 625 h, respectively.

11.8 Model VII

This mathematical model represents a two-identical redundant active unit system used in the oil and gas industry subjected to human errors [3,4]. Each unit can fail either due to a hardware failure or due to a human error. For the successful operation of the oil and gas industry system, at least one unit must function normally.

The state space diagram for the oil and gas industry system is shown in Figure 11.6. The numerals and small single letters in the boxes and circles in Figure 11.6 denote system states.

The model is subjected to the following assumptions:

- Hardware failures and human errors occur independently.
- Both units of the system are identical and operate simultaneously.
- Failures of each unit can be classified under two categories: failures due to hardware problems and failures due to human errors.
- Each unit can fail either due to a human error or hardware failure.
- Hardware failure and human error rates are constant.

The following symbols are associated with the model:

i is the ith state of the oil and gas industry system, where $i = 0$ means that both units of the oil and gas industry system are operating normally, $i = 1$ means that one unit failed due to a hardware failure and the other is working normally, $i = 2$ means that one unit failed due to a human error and the other is working normally, $i = h$ means that both units failed due to hardware failures, and $i = e$ means that both units failed due to human errors.

$P_i(t)$ is the probability that the oil and gas industry system is in state i at time t, for $i = 0, 1, 2, h, e$.

λ is the constant hardware failure rate of a unit.

γ is the constant human error rate of a unit.

Using the Markov method described in Chapter 4, we write down the following system of differential equations for Figure 11.6 [3,4]:

$$\frac{dP_0(t)}{dt} + (2\lambda + 2\gamma)P_0(t) = 0 \tag{11.56}$$

$$\frac{dP_1(t)}{dt} + (\lambda + \gamma)P_1(t) = 2\lambda P_0(t) \tag{11.57}$$

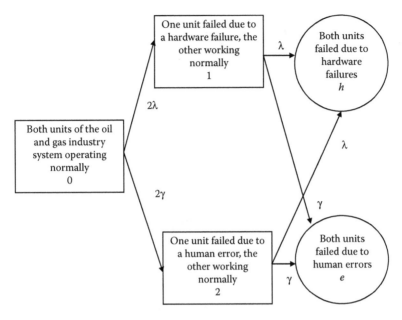

FIGURE 11.6
State space diagram of a two-identical redundant active unit oil and gas industry system.

$$\frac{dP_2(t)}{dt} + (\lambda + \gamma)P_2(t) = 2\gamma P_0(t) \tag{11.58}$$

$$\frac{dP_h(t)}{dt} = \lambda P_1(t) + \lambda P_2(t) \tag{11.59}$$

$$\frac{dP_e(t)}{dt} = \gamma P_1(t) + \gamma P_2(t) \tag{11.60}$$

At time $t = 0$, $P_0(0) = 1$, and $P_1(0) = P_2(0) = P_h(0) = P_e(0) = 0$.

By solving Equations 11.56 through 11.60, we get the following equations for the oil and gas industry system state probabilities at time t:

$$P_0(t) = e^{-2(\lambda + \gamma)t} \tag{11.61}$$

$$P_1(t) = \frac{2\lambda}{(\lambda + \gamma)}[e^{-(\lambda + \gamma)t} - e^{-2(\lambda + \gamma)t}] \tag{11.62}$$

$$P_2(t) = \frac{2\gamma}{(\lambda+\gamma)} [e^{-(\lambda+\gamma)t} - e^{-2(\lambda+\gamma)t}] \qquad (11.63)$$

$$P_h(t) = \frac{\lambda}{(\lambda+\gamma)} [1 - e^{-(\lambda+\gamma)t}]^2 \qquad (11.64)$$

$$P_e(t) = \frac{\gamma}{(\lambda+\gamma)} [1 - e^{-(\lambda+\gamma)t}]^2 \qquad (11.65)$$

The oil and gas industry system reliability is given by

$$\begin{aligned}
R_{os}(t) &= P_0(t) + P_1(t) + P_2(t) \\
&= 1 - [1 - e^{-(\lambda+\gamma)t}]^2
\end{aligned} \qquad (11.66)$$

where
$R_{os}(t)$ is the reliability of the oil and gas industry system at time t.

By integrating Equation 11.66 over the time interval [0,∞], we get [14]

$$\begin{aligned}
\text{MTTF}_{os} &= \int_0^\infty R_{0s}(t)\,dt \\
&= \frac{3}{2(\lambda+\gamma)}
\end{aligned} \qquad (11.67)$$

where
MTTF_{os} is the mean time to failure of a two-redundant active unit oil and gas industry system subjected to human errors.

EXAMPLE 11.6

An oil and gas industry system is composed of two active, identical, and independent units. At least one unit must operate normally for its (i.e., system) successful operation. Each unit can fail either due to a hardware failure or due to a human error. A unit's constant hardware failure and human error rates are 0.009 failures/h and 0.002 errors/h, respectively.

Calculate the oil and gas industry system reliability for a 10-h mission and mean time to failure.

By inserting the given data values into Equation 11.66, we get

$$R_{os}(10) = 1 - [1 - e^{-(0.009+0.002)(10)}]^2$$

$$= 0.9891$$

Similarly, by inserting the given data values into Equation 11.67, we get

$$\text{MTTF}_{os} = \frac{3}{2(0.009 + 0.002)}$$

$$= 136.36\,\text{h}$$

Thus, the oil and gas industry system reliability and mean time to failure for the specified data values are 0.9891 and 136.36 h, respectively.

PROBLEMS

1. Assume that an oil and gas industry worker is performing a time-continuous task and his/her error rate is 0.008 errors/h. Calculate the reliability of the worker for an 8-h mission.

2. Write an essay on mathematical models used for performing reliability and safety-related analysis in the oil and gas industrial sector.

3. Prove Equations 11.4 through 11.6 by using Equations 11.1 through 11.3.

4. A system used in the oil and gas industry can fail safely or unsafely, and its constant failure rates are 0.0006 and 0.0001 failures/h, respectively. Calculate the probabilities of the system failing safely and unsafely during a 100-h mission.

5. Prove Equations 11.12 through 11.14 by using Equations 11.9 through 11.11.

6. Assume that a worker is performing an oil and gas industry-related task in a fluctuating environment (i.e., normal and stressful) and his/her constant error rates are 0.002 and 0.0007 errors/h, respectively. The constant transition rates from normal to stressful environment and from stressful to normal environment are 0.002 per hour and 0.001 per hour, respectively. Calculate his/her mean time to error.

7. Prove Equations 11.36 through 11.37 by using Equations 11.32 through 11.35.

8. Prove Equation 11.30 by using Equations 11.25 and 11.26.

9. Assume that in Figure 11.5 for transition rates λ_1, λ_2, λ_3, and λ_4, we have the following given values:

   - $\lambda_1 = 0.0005$ failures/h
   - $\lambda_2 = 0.0001$ failures/h
   - $\lambda_3 = 0.0004$ failures/h
   - $\lambda_4 = 0.0002$ failures/h

   Calculate the oil and gas industry system reliability for a 15-h mission and mean time to failure.

10. Prove Equation 11.67 by using Equations 11.61 through 11.63.

References

1. Regulinski, T.L., Asksren, W.B., Stochastic modeling of human performance effectiveness functions, *Proceedings of the Annual Reliability and Maintainability Symposium*, 1972, pp. 407–416.
2. Shooman, M.L., *Probabilistic Reliability: An Engineering Approach*, McGraw-Hill Book Company, New York, 1968.
3. Dhillon, B.S., *Human Reliability: With Human Factors*, Pergamon Press, New York, 1986.
4. Dhillon, B.S., *Human Reliability, Error, and Human Factors in Power Generation*, Springer, Inc., London, 2014.
5. Dhillon, B.S., *Human Reliability, Error, and Human Factors in Engineering Maintenance: With Reference to Aviation and Power Generation*, CRC Press, Boca Raton, Florida, 2009.
6. Dhillon, B.S., *Human Reliability and Error in Medical System*, World Scientific Publishing, River Edge, New York, 2003.
7. Dhillon, B.S., *Human Reliability and Error in Transportation Systems*, Springer, Inc., London, 2007.
8. Dhillon, B.S., *Mine Safety: A Modern Approach*, Springer, Inc., London, 2010.
9. Dhillon, B.S., The Analysis of the Reliability of Multi-State Device Networks, Doctoral Dissertation, 1975. Available from the National Library of Canada, Ottawa, Canada.
10. Dhillon, B.S., Stochastic models for predicting human reliability, *Microelectronics and Reliability*, 25, 1985, 729–752.
11. Reglinski, T.L., Askren, W.B., Mathematical modeling of human performance reliability, *Proceedings of the Annual Symposium on Reliability*, 1969, pp. 5–11.
12. Askren, W.B., Regulinski, T.L., Quantifying human performance for reliability analysis of systems, *Human Factors*, 11, 1969, 393–396.
13. Dhillon, B.S., Rayapati, S.N., Reliability and availability analysis of surface transit systems, *Microelectronics and Reliability*, 24, 1984, 1029–1033.
14. Dhillon, B.S., *Design Reliability: Fundamentals and Applications*, CRC Press, Boca Raton, Florida, 1999.
15. Dhillon, B.S., Rayapati, S.N., Reliability evaluation of transportation systems with human errors, *Proceedings of the IASTED International Conference on Applied Simulation and Modeling*, 1985, pp. 4–7.

Bibliography: Literature on Safety and Reliability in the Oil and Gas Industry

A.1 Introduction

Many publications on safety and reliability in the oil and gas industry have appeared over the years in the form of journal articles, conference proceedings articles, technical reports, and so on. This appendix presents an extensive list of selective publications related, directly or indirectly, to safety and reliability in the oil and gas industry. The period covered by the listing is from 1928 to 2014. The objective of this listing is to provide readers with sources for obtaining additional information on safety and reliability in the oil and gas industry.

A.2 Publications

Aas, A.L., The Human Factors Assessment and Classification System (HFACS) for the oil gas industry, *International Petroleum Technology Conference*, 2008, 2325–2335.

Al-Dhafiri, S. et al., Sharing experience of high corrosion rate in acid gas removal plant, *Proceedings of the Conference on Corrosion*, 2012, pp. 1993–2004.

Alfon, P. et al., Pipeline material reliability analysis regarding to probability of failure using corrosion degradation model, *Proceedings of the 2nd International Conference on Advances in Materials and Manufacturing Processes*, 2012, pp. 705–715.

Al-Qahtani, A.M., Lourido, M.L., Dabbousi, R.M.O., Al-Shahrani, O.O., Management of electrical equipment obsolescence at oil gas industrial facilities, *Proceedings of the 57th Annual Petroleum and Chemical Industry Conference*, 2010, pp. 10–15.

Al-Qahtani, A.M., Lourido, M.L., Dabbousi, R.M.O., Al-Shahrani, O.O., Obsolescence of aging electrical equipment: Management program in oil and gas industrial facilities, *IEEE Industry Applications Magazine*, 18, 2012, 37–45.

Amaral, S.P., Rodrigues, E.C., Carvalho Filho, A.F., Health, safety and environment management assessment tool: Application in the Brazilian oil industry, *Proceedings of the SPE International Conference on Health, Safety and Environment in Oil and Gas Exploration and Production*, 2004, pp. 611–616.

Antonsen, S., Skarholt, K., Ringstad, A.J., The role of standardization in safety management: A case study of a major oil gas company, *Safety Science*, 50, 2012, 2001–2009.

Aven, T., Vinnem, J.E., Wiencke, H.S., A decision framework for risk management, with application to the offshore oil and gas industry, *Reliability Engineering and System Safety*, 92, 2007, 433–448.

Backus, O., Fuel oil storage effect and impact on utility's gas turbine operation, *Proceedings of the International Gas Turbine and Aero-engine Congress and
Exposition*, 1991, pp. 10–15.

Balderas, J., Gelpke, D., Salawage, C., Properly designed PPE for the oil and gas service industry, *Proceedings of the SPE International Conference on Health, Safety, and
Environment in Oil and Gas Exploration and Production*, 2010, pp. 644–649.

Barabadi, A., Barabady, J., Markeset, T., Application of accelerated failure model for
the oil and gas industry in the Arctic region, *Proceedings of the IEEE International
Conference on Industrial Engineering and Engineering Management*, 2010, pp.
2244–2248.

Barabadi, A., Markeset, T., Reliability and maintainability performance under
Arctic conditions, *International Journal of Systems Assurance Engineering and
Management*, 2, 2011, 205–217.

Barbosa, C. et al., Failure analysis of two stainless steel based components used in an
oil refinery, *Journal of Failure Analysis and Prevention*, 8, 2008, 320–326.

Barling, J., Dupre, K.E., Kavanagh, J.M., Kenny, S., Khan, F., Veitch, B., The integration of high performance work systems and workplace safety in oil and gas
industry, *Proceedings of the 9th International Conference on Health, Safety, and
Environment in Oil and Gas Exploration and Production*, 2008, pp. 526–537.

Barrell, A.C., Sharp, J.V., Safety research in the offshore oil and gas industry, *Process
Safety and Environmental Protection: Transactions of the Institution of Chemical
Engineers, Part B*, 72, 1994, 229–233.

Basrawi, M., Keck, D., Non-destructive testing technologies for the oil industry, *Proceedings of the 13th Middle East Oil and Gas Show and Conference*, 2003,
pp. 1017–1021.

Bhatnagar, V., Smith, A., Berger, S., Managing process safety of the upstream sector:
Lessons learnt from the downstream industry, *Proceedings of the Offshore Europe
Oil and Gas Conference*, 2011, pp. 971–977.

Bradish, R., Bui, J., Nunez, Y., Global challenges facing the health, safety, and environmental (HSE) function within the oil and gas industry, *Proceedings of the SPE
Annual Technical Conference and Exhibition*, 2008, pp. 2158–2161.

Brewerton, R.W. et al., Experience in the application of the goal setting regime
and risk-based design of safety critical elements in the oil and gas industry,
*Proceedings of the ASME 32nd International Conference on Ocean, Offshore, and
Arctic Engineering*, 2013, pp. 52–57.

Calixto, E., The safety integrity level as Hazop risk consistence: The Brazilian risk
analysis study case, *Proceedings of the European Safety and Reliability Conference*,
2007, pp. 629–634.

Campbell III, H.H., Oil industry challenges for the 21st century, *Welding Journal*, 89,
2010, 50–52.

Cardoso, A.G., Do Sameiro Queiros, M., Meireles, J.M., Diogo, M.T., Safety management on a refinery maintenance turnaround, *Proceedings of the 10th International
Symposium on Occupational Safety and Hygiene*, 2014, pp. 505–510.

Catelani, M., Ciani, L., Luongo, V., Safety analysis in oil gas industry in compliance
with standards IEC61508 and IEC61511: Methods and applications, *Proceedings
of the IEEE International Instrumentation and Measurement Technology Conference*,
2013, pp. 686–690.

Cavaney, R., US oil, gas industry enhances health, safety, and environment, *Oil and
Gas Journal*, 98, 2000, 64–66, 68–69.

Chabot, W., Oil spill prevention on the California coast: Facts, failures, and the future, *Proceedings of the Conference on California and the World Ocean*, 1997, pp. 563–564.

Chakraborty, A.B., Carbon management: The emerging paradigm for the oil industry, *Proceedings of the SPE Annual Technical Conference*, 2007, pp. 2410–2415.

Champlan, D., Development of offshore oil deposits: Production and safety, *Revue de L'Institut Francais du Petrole*, 39, 1984, 267–289.

Chauvin, D., Depraz, S., Buckley, H., Saving energy in the oil and gas industry, *Proceedings of the 9th International Conference on Health, Safety, and Environment in Oil and Gas Exploration and Production*, 2008, pp. 1881–1890.

Chen, X., Ai, Z., Fan, Z., Hu, J., Guan, W., Cheng, C., Accident investigation and risk assessment of Chinese Industrial Pipelines, *Proceedings of the ASME Pressure Vessels and Piping Conference*, 2009, pp. 1177–1186.

Chis, T., Pipeline accident statistics base to pipeline rehabilitation, *Proceedings of the International Pipeline Conference*, 1996, pp. 319–327.

Chmilar, W.S., Pipeline reliability evaluation and design using the sustainable capacity methodology, *Proceedings of the 1st International Pipeline Conference*, 1996, pp. 803–810.

Clark, E., Edwards, C., Perry, P., Campbell, G., Stevens, M., Helicopter safety in the oil and gas business, *Proceedings of the IADC/SPE Drilling Conference*, 2006, pp. 211–249.

Cloughley, T.M.G., Thomas, I., Accident data for the upstream oil gas industry, *Proceedings of the 3rd International Conference on Health, Safety, and Environment in Oil and Gas Exploration and Production*, 1996, pp. 161–169.

Consiglio, M. et al., A guide to social impact assessment in the oil and gas industry, *Proceeding of the 8th SPE International Conference on Health, Safety, and Environmental in Oil and Gas Exploration and Production*, 2006, pp. 390–395.

Crichton, M., Attitudes to teamwork, leadership, and stress in oil drilling teams, *Safety Science*, 43, 2005, 679–696.

Crutcher, R.A., Yip, R.Y., Young, M.R., Employee assistance programs in the upstream petroleum industry: A Canadian perspective, *Proceedings of the First International Conference on Health, Safety, and Environment in Oil and Gas Exploration and Production*, 1991, pp. 753–759.

Davies, S., Deep oil dilemma: Explosion and sinking of Deepwater Horizon, *Engineering and Technology*, 5, 2010, 44–49.

De, P.R.H., Whiddon, D.J., Strategy of remedial investigations, *Proceedings of the 2nd International Conference on Health, Safety, and Environment in Oil and Gas Exploration and Production*, 1994, pp. 649–659.

De Morais, C.P.M., Application of the 17 practices of the management system for operational safety on marine installations for drilling and production of oil and natural gas in Brazil, *Proceedings of the Offshore Technology Conference*, 2011, pp. 1164–1190.

Dell, J.J., Meakin, S., Cramwinckel, J., Sustainable water management in the oil and gas industry: Use of the WBCSD global water tool to map risks, *Proceedings of the 9th International Conference on Health, Safety, and Environment in Oil and Gas Exploration and Production*, 2008, pp. 2004–2017.

Den Haan, K., Downstream oil industry safety statistics for 2010, *CONCAWE Review*, 20, 2011, 16–18.

Denney, D., Strategic direction for reducing fatal oil and gas industry incidents, *Journal of Petroleum Technology*, 57, 2005, 66–68.

Dey, P.K., Gupta, S.S., Ho, W., Managing technology in oil pipelines industry, *International Journal of Services, Technology, and Management*, 7, 2006, 185–210.

Didla, S., Mearns, K., Flin, R., Safety citizenship behaviour in the oil and gas industry, *Proceedings of the European Safety and Reliability Conference*, 2007, pp. 2451–2456.

Doherty, B.D., Fragu, L.P., Sustainable HSE performance: Successful management systems and monitoring tools in the middle east LNG industry, *Proceedings of the SPE International Conference on Health, Safety, and Environment in Oil and Gas Exploration and Production*, 2010, pp. 1445–1463.

Dowd, D., Daher, E., Safety KPIs during shutdown turnaround—What to measure and how to impact the overall economics, *Proceedings of the SPE Middle East Health, Safety, Security, and Environment Conference and Exhibition*, 2012, pp. 138–145.

Drakeley, B.K., Douglas, N.I., Haugen, K.E., Willmann, E., Application of reliability analysis techniques to intelligent wells, *SPE Drilling and Completion*, 18, 2003, 159–168.

Duffuaa, S.O., Daya, M.A.B., Turnaround maintenance in petrochemical industry: Practices and suggested improvements, *Journal of Quality in Maintenance Engineering*, 10, 2004, 184–190.

Dursun, S., Rangarajan, K., Singh, A., A framework-oriented approach for determining attribute importance when building effective predictive models for oil and gas data analytics, *Proceedings of the SPE Annual Technical Conference and Exhibition*, 2013, pp. 2670–2686.

Emslie, D.A. et al., Safety management in the bass strait operations, *Proceedings of the SPE Asia-Pacific Conference*, 1991, pp. 677–683.

Fabiano, B., Curro, F., From a survey on accidents in the downstream oil industry to the development of a detailed near-miss reporting system, *Process Safety and Environmental Protection*, 90, 2012, 357–367.

Falker, III J.M., Nickerson, W., A new direction for safety policy: The offshore oil industry and safety regulation of technology, *Technology in Society*, 18, 1996, 503–510.

Fan, Z. et al., The elemental sulfur deposition and its corrosion in high sulfur gas fields, *Natural Gas Industry*, 33, 2013, 102–109.

Fei, L.W., Applicability of dynamic risk assessment in the oil and gas industry, *Proceedings of the SPE Middle East Health, Safety, Security, and Environment Conference*, 2012, pp. 70–76.

Fennessey, K.D., Romer, R.F., Dell, J.J., New tools to assess water risk and enhance environmental and operational performance in oil and gas, *Proceedings of the SPE/APPEA International Conference on Health, Safety, and Environment in Oil and Gas Exploration and Production*, 2012, pp. 2380–2383.

Ferreira, L.A., Gaspar, D., Silva, J.L., Failure data analysis of an oil refinery centrifugal pumps, *Proceedings of the 11th International Probabilistic Safety Assessment and Management Conference and the Annual European Safety and Reliability Conference*, 2012, pp. 1422–1430.

Fleming, M., Effective supervisory safety leadership behaviors in the oil and gas industry, *Proceedings of the HAZARDS XV: The Process, Its Safety, and the Environment "Getting It Right" Conference*, 2000, pp. 371–384.

Flin, R. et al., Risk perception by offshore workers on UK oil and gas platforms, *Safety Science*, 22, 1996, 131–145.

Flin, R.H. et al., Risk perception and safety in the UK offshore oil and gas industry, *Proceedings of the 3rd International Conference on Health, Safety, and Environment in Oil and Gas Exploration and Production*, 1996, pp. 187–197.

Flynn, L., Kaitano, A.E., Bery, R., Emerging pandemic threats and the oil and gas industry, *Proceedings of SPE/APPEA International Conference on Health, Safety, and Environment in Oil and Gas Exploration and Production*, 2012, pp. 316–325.

Fu, J. et al., Environmental and safety risk management of oil and gas pipelines in their full life cycle, *Natural Gas Industry*, 33, 2013, 138–143.

Gill, C., The role of geophysical technology in reducing risk for oil and gas operations, *Proceedings of the 73rd European Association of Geoscientists and Engineers Conference and Exhibition*, 2011, pp. 1011–1013.

Gordon, R.P.E., Contribution of human factors to accidents in the offshore oil industry, *Reliability Engineering and System Safety*, 61, 1998, 95–108.

Gordon, R.P.E., Flin, R.H., Mearns, K., Fleming, M.T., Assessing the human factors causes of accidents in offshore oil industry, *Proceedings of the 3rd International Conference on Health, Safety, and Environment in Oil and Gas Exploration and Production*, 1996, pp. 635–644.

Gould, K.S., Ringstad, A.J., Van De Merwe, K., Human reliability analysis in major accident risk analyses in the Norwegian petroleum industry, *Proceedings of the Human Factors and Ergonomics Society 56th Annual Meeting*, 2012, pp. 2016–2020.

Goyal, R.K., Probabilistic risk analysis: Two case studies from the oil industry, *Professional Safety*, 31, 1986, 11–19.

Gran, B.A., Nyheim, O.M., Seljelid, J., Vinnem, J.E., A BBN risk model of maintenance work on major process equipment on offshore petroleum installations, *Proceedings of the European Safety and Reliability Conference*, 2012, pp. 1249–1257.

Grice, K.J., Naturally occurring radioactive materials (NORM) in the oil and gas industry: A new challenge, *Proceedings of the First International Conference on Health, Safety, and Environment in Oil and Gas Exploration and Production*, 1991, pp. 559–571.

Hansson, L., Lamvik, G.M., Antonsen, S., Drilling consortia—New ways of organising exploration drilling in the oil and gas industry and the consequences for safety, *Proceedings of the European Safety and Reliability Conference*, 2012, pp. 1701–1708.

Hauge, S., Lundteigen, M.A., A new approach for follow-up of safety instrumented systems in the oil and gas industry, *Proceedings of the Joint European Safety and Reliability and Society for Risk Analysis Europe Conference*, 2009, pp. 2921–2928.

Hauge, S., Lundteigen, M.A., Hokstad, P., Habrekke, S., Reliability prediction method for safety instrumented systems-PDS method, *SINTEF Report STF50A*, 6031, 2010, 1–50.

Hendrex, T., Oil industry launches new marine terminal information system, *Materials Performance*, 52, 2013, 29–31.

Herrmann, H.J., Paul, K.D., Management of safety in oil industry, *Chemical Engineering and Technology*, 32, 2009, 199–206.

Hidalgo, E.M.P., Silva, D.W.R., De Souza, G.F.M., Application of Markov chain to determine the electric energy supply system reliability for the cargo control system of LNG carriers, *Proceedings of the 32nd International Conference on Ocean, Offshore, and Arctic Engineering*, 2013, pp. 200–209.

Hightower, C.L., "Safety first" in oil industry, *Petroleum World*, 31, 1934, 16–19.

Hokstad, P., Vatn, J., Aven, T., Sorum, M., Use of risk acceptance criteria in Norwegian offshore industry: Dilemmas and challenges, *Risk, Decision and Policy*, 9, 2004, 193–206.

Hopkins, J., The practical application of process safety to determine and monitor asset integrity of oil and gas facilities, *Proceedings of the Offshore Europe and Gas Conference*, 2011, pp. 868–901.

Hopkins, J., Essential process safety management for managing multiple oil and gas assets, *Proceedings of the 23rd Institution of Chemical Engineers Symposium on Hazards*, 2012, pp. 409–416.

Hovden, J., Lie, T., Karlsen, J.E., Alteren, B., The safety representative under pressure: A study of occupational health and safety management in the Norwegian oil and gas industry, *Safety Science*, 46, 2008, 493–509.

Hu, J.Q., Zhang, L.B., Wang, Z.H., Linag, W., The application of integrated diagnosis database technology in safety management of oil pipeline and transferring pump units, *Journal of Loss Prevention in the Process Industries*, 22, 2009, 1025–1033.

Ikeagwuani, U.M., John, G.A., Safety in maritime oil sector: Content analysis of machinery space fire hazards, *Safety Science*, 51, 2013, 347–353.

Isaksen, S., Anbalagan, A., A study of factors which influence the failure rate of equipment in the oil and gas industry, *Proceedings of the European Safety and Reliability Annual Conference*, 2010, pp. 57–64.

Jacobson, E., Spector, Y., Monitoring high risk industries with open path optical gas detection systems, *Proceedings of the Waste Management Association's 89th Annual Meeting*, 1996, pp. 12–14.

Janus, B., Murphy, H., Sustainability reporting and the oil and gas industry: Challenges and emerging trends, *Proceedings of the SPE European HSE Conference and Exhibition*, 2013, pp. 183–189.

Jdestl, K.A. et al., Achieving an industry standard in the assessment of environmental risk: Oil spill risk management and mira method, *Proceedings of the International Oil Spill Conference*, 2005, pp. 81–91.

Jiang, D., Management system of safety and environmental protection in China offshore oil industry, *Proceedings of the 6th International Oil and Gas Conference and Exhibition in China*, 1998, pp. 13–20.

Johnson, O.W., Oil industry's interest in fire safety, *National Fire Protection Association Quarterly*, 44, 1950, 21–25.

Kermani, M.B., Harrop, D., Impact of corrosion on oil and gas industry, *Proceedings of the 9th Middle East Oil Conference*, 1995, pp. 135–148.

Kermani, M.B., Harrop, D., Impact of corrosion on oil and gas industry, *SPE Production and Facilities*, 11, 1996, 186–190.

Khalsa, A.S., Social risk management cheat sheet: Management strategies for the oil and gas industry, *Proceedings of the Latin American and Caribbean Health, Safety, Environment, and Social Responsibility Conference*, 2013, pp. 198–203.

Kitwik, U., Ai-Yakoot, S., Al-Ali, F., Implementing process safety—Kuwait Petroleum Corp., *Proceedings of the 23rd Center for Chemical Process Safety International Conference*, 2008, pp. 72–94.

Knochenhauer, M., Bakken, B.I., Baas, T.L, Process safety, instrumented safety barriers—what can we learn from the nuclear industry?, *Proceedings of the PCIC Europe Conference*, 2010, pp. 1–6.

Kong, X., Ohadi, M.M., Applications of micro and nano technologies in the oil and gas industry: A review of the recent progress, *Proceedings of the 14th Abu Dhabi International Petroleum Exhibition and Conference*, 2010, pp. 1703–1713.

Kumarasamy, D., Karthikeyan, S., Haze hazard and emergency in the oil and gas industry, *Proceedings of the 9th International Conference on Health, Safety, and Environment in Oil and Gas Exploration and Production*, 2008, pp. 486–508.

Landucci, G. et al., Hazard assessment of edible oil refining: Formation of flammable mixtures in storage tanks, *Journal of Food Engineering*, 105, 2011, 105–111.

Lawrie, G., Samoylova, O., Health, safety environment in the Russian oil field, *Proceedings of the SPE International Conference on Health, Safety, and Environment in Oil and Gas Exploration and Production*, 2010, pp. 2260–2264.

Lee, T.W., Othman, N.S., Managing environmental incidents at oil and gas facilities before they happen, *Proceedings of the 9th International Conference on Health, Safety, and Environment in Oil and Gas Exploration and Production*, 2008, pp. 692–699.

Lins, I.D. et al., Reliability prediction of oil production wells by particle swarm optimized support vector machines, *Proceedings of the European Safety and Reliability Annual Conference*, 2010, pp. 914–922.

Luna-Mejias, G., Process safety management—A practical view, *Proceedings of the 28th Center for Chemical Process Safety International Conference*, 2013, pp. 190–201.

Lundteigen, M.A., Rausand, M., Common cause failures in safety instrumented systems on oil and gas installations: Implementing defense measures through function testing, *Journal of Loss Prevention in the Process Industries*, 20, 2007, 218–229.

Lundteigen, M.A., Rausand, M., Spurious activation of safety instrumented systems in the oil and gas industry: Basic concepts and formulas, *Reliability Engineering and System Safety*, 93, 2008, 1208–1217.

Lundteigen, M.A., Rausand, M., Reliability assessment of safety instrumented systems in the oil and gas industry: A practical approach and a case study, *International Journal of Reliability, Quality, and Safety Engineering*, 16, 2009, 187–212.

Lynch, M.C., The Macondo oil field disaster, *Journal of Disaster Research*, 6, 2011, 482–485.

Lyra Da Silva, B.B. et al., Helicopter offshore safety in the Brazilian oil and gas industry, *Proceedings of the IEEE Systems and Information Engineering Design Symposium*, 2005, pp. 235–241.

Mackie, S.I., Begg, S.H., Smith, C., Welsh, M.B., Human decision making in the oil and gas industry, *Proceedings of the SPE Asia Pacific Oil and Gas Conference and Exhibition*, 2010, pp. 117–124.

Magnussen, B.F. et al., Computational analysis of large-scale fires in complex geometrics—A means to safeguard people and structural integrity in the oil and gas industry, *Chemical Engineering Transactions*, 31, 2013, 793–798.

Maharaj, P.S., Dyal, S., Ramnath, K., Health, safety, and environmental management systems auditing for an integrated oil and gas company in Trinidad and Tobago, *Proceedings of the Engineering Technology Conference on Energy*, 2002, pp. 429–434.

Mannan, M.S., Olewski, T., Waldram, S., Safety in the oil and gas industries in Qatar, *Proceedings of the 21st Institution of Chemical Engineers Symposium on Hazards*, 2009, pp. 570–575.

Marquez, C.P. et al., Reliability stochastic model applied to evaluate the economic impact of the failure in the life cycle cost analysis (LCCA): Case of study in the oil industry, *Proceedings of the European Safety and Reliability Annual Conference,* 2010, pp. 625–637.

Martin, A., Health performance in the oil and gas industry—The results, *Proceedings of the SPE/APPEA International Conference on Health, Safety, and Environment in Oil and Gas Exploration and Production,* 2012, pp. 196–205.

Martins, M.R., Maturana, M.C., Human error contribution in collision and grounding of oil tankers, *Risk Analysis,* 30, 2010, 674–698.

Mearns, K., Whitaker, S.M., Flin, R., Safety climate, safety management practice and safety performance in offshore environments, *Safety Science,* 41, 2003, 641–680.

Mearns, K., Yule, S., The role of national culture in determining safety performance: Challenges for the global oil and gas industry, *Safety Science,* 47, 2009, 777–785.

Merritt, S., The impacts of national culture on safety culture in the global oil and gas industry, *Proceedings of the SPE/APPEA International Conference on Health, Safety, and Environment in Oil and Gas Exploration and Production,* 2012, pp. 594–605.

Mika, F. et al., Stress and social anxiety assessment among offshore personnel in oil and gas industry, *Proceedings of the International Conference on Health, Safety, and Environment in Oil and Gas Exploration and Production,* 2012, pp. 399–405.

Mika, F., Martin, A., Cox, R., Effective tuberculosis management in the oil gas industry, *Proceedings of the SPE Middle East Health, Safety, Security, and Environment Conference and Exhibition,* 2010, pp. 382–388.

Mohamed, H., Donnelly, R., Fraser, A., Fitness to work: A risk-based approach for the oil and gas industry, *Proceedings of the SPE/APPEA International Conference on Health, Safety, and Environment in Oil and Gas Exploration and Production,* 2012, pp. 1629–1642.

Moura, M.C. et al., Predictive maintenance policy for oil well equipment in the case scaling through support vector machines, *Proceedings of the European Safety and Reliability Conference: Advances in Safety, Reliability, and Risk Management,* 2012, pp. 503–507.

Muehlenbachs, L., Cohen, M.A., Gerarden, T., The impact of water depth on safety and environmental performance in offshore oil and gas production, *Energy Policy,* 55, 2013, 699–705.

Muir, J., Reidinger, R., Chan, Y.M., Capturing sustainability issues in the oil and gas industry, *Proceedings of the SPE International Conference on Health, Safety, and Environment in Oil and Gas Exploration and Production,* 2002, pp. 718–724.

Naidoo, R., Manning, E.J., Optimal network topology and reliability Indices to be used in the design of power distribution networks in oil and gas plants, *Australian Journal of Electrical and Electronics Engineering,* 10, 2013, 239–250.

Nicholas, L., Lozier, W., Case studies demonstrating sustainability and risk evaluations in environmental due diligence for upstream oil and gas transactions in Alberta, *Proceedings of the SPE Americas E and P Health, Safety, Security, and Environmental Conference,* 2013, pp. 359–366.

O'Dea, A., Flin, R., Site managers and safety leadership in the offshore oil and gas industry, *Safety Science,* 37, 2001, 39–57.

Okoh, P., Haugen, S., The influence of maintenance on some selected major accidents, *Chemical Engineering Transactions,* 31, 2013, 493–498.

Olsen, E., Exploring the possibility of a common structural model measuring associations between safety climate factors and safety behaviour in health care and petroleum sectors, *Accident Analysis and Prevention*, 42, 2010, 1507–1516.

Ontko, R.J., Bradley III, D.D., The great emissions roundup: Strategies for permitting maintenance, start up, and shutdown (MSS) emissions at upstream oil gas facilities, *Proceedings of the SPE Americas E and P Health, Safety, Security, and Environmental Conference*, 2013, pp. 234–238.

Ormond, A., De Klerk, M., Implementing improvements in process safety in Corus, *Proceedings of the 21st Institution of Chemical Engineers Symposium on Hazards*, 2009, pp. 576–586.

Parkes, K., Hodkiewicz, M., Morrison D., The role of organizational factors in achieving reliability in the design and manufacture of subsea equipment, *Human Factors and Ergonomics in Manufacturing*, 22, 2012, 487–505.

Pedroni, P.M. et al., Integrating the emerging concepts of ecosystem services into oil and gas industry environmental management practices, *Proceedings of the SPE/ APPEA International Conference on Health, Safety, and Environment in Oil and Gas Exploration and Production*, 2012, pp. 98–111.

Peng, X.Y. et al., Oil/gas pipeline typical third-party endanger reliability research considering failure transfer/model correlation, *Proceedings of the International Conference on Pipelines and Trenchless Technology*, 2011, pp. 1728–1739.

Planeix, M. et al., SAIPEM's leadership in safety program: Bringing a new safety culture to the oil gas construction industry, *Proceedings of the 9th International Conference on Health, Safety, and Environment in Oil and Gas Exploration and Production*, 2008, pp. 153–164.

Polyanskii, A.M., Polyanskii, V.A., Yakovlev, Y.A., Metrological assurance of measurement of hydrogen concentration in metals—A basis for safety in the oil and gas industry, *Measurement Techniques*, 56, 2013, 328–333.

Powell, N., Owen, D., Practical application of competence management systems in safety critical industries, *Proceedings of the International Conference on Contemporary Ergonomics and Human Factors*, 2011, pp. 429–435.

Price, J.C., Fitness-for-purpose failure and corrosion control management in offshore oil and gas development, *Proceedings of the 11th International Offshore and Polar Engineering Conference*, 2001, pp. 234–241.

Rahimi, M., Rausand, M., Wu, S., Reliability prediction of offshore oil and gas equipment for use in an arctic environment, *Proceedings of the International Conference on Quality, Reliability, Risk, Maintenance, and Safety Engineering*, 2011, pp. 81–86.

Rains, B., Increasing the agility of process safety management systems, *Proceedings of the Abu Dhabi International Petroleum Exhibition and Conference*, 2012, pp. 2091–2098.

Rakin, M. et al., Structural integrity assurance of casing pipes in the oil and gas industry, *Proceedings of the 5th International Conference on Safety and Security Engineering*, 2013, pp. 401–410.

Ratliff, M., Changing safety paradigms in the oil and gas industry, *Proceedings of the SPE Annual Technical Conference and Exhibition*, 2004, pp. 3981–3986.

Ravi, K., Bosma, M., Gastebled, O., Improve the economics of oil and gas wells by reducing the risk of cement failure, *Proceedings of the IACD/SPE Drilling Conference*, 2002, pp. 377–389.

Raza, J., Liyanage, J.P., Technical integrity and performance optimization for enhanced reliability in "Smart Assets"; Case of a North Sea oil gas production facility, *Proceedings of the European Safety and Reliability Conference*, 2007, pp. 2509–2516.

Red, W., Vinnem, J.E., Evaluation of accidents and incidents in the offshore oil and gas industry by use of safety barrier performance diagrams, *Proceedings of the European Safety and Reliability Conference*, 2006, pp. 63–69.

Renforth, L. et al. Continuous, remote on-line partial discharge (OLPD) monitoring of HV EX/ATEX motors in the oil and gas industry, *Proceedings of the 60th Annual IEEE Petroleum and Chemical Industry Technical Conference*, 2013, pp. 20–25.

Richmond, J.E., Application of reliability engineering in the oil and gas industry, *Proceedings of the Annual Reliability and Maintainability Symposium*, 1983, pp. 491–493.

Ritwik, U., Ai-Yakoot, S., Al-Ali, F., Implementing process safety—Kuwait Petroleum Corp., *Proceedings of the 23rd Center for Chemical Process Safety International Conference*, 2008, pp. 72–94.

Rizwan, M., Al-Marri, H., Safety management in oil gas industry—The how's and the why's, *Proceedings of the SPE Production and Operation Symposium*, 2012, pp. 399–410.

Ruvalcaba Velarde, S.E., Al-Ghamdi, A.A., The impact of an infrastructure reliability and data communication index on improving intelligent field operations in an oilfield in Saudi Arabia, *Proceedings of the 18th Middle East Oil and Gas Show and Conference*, 2013, pp. 1853–1868.

Salvador, A. et al., Risk management and prevention program to avoid loss of process (LP3) based on human reliability, practical application and measurement, *Proceedings of the AICHE Spring Meeting and 7th Global Congress on Process Safety*, 2011, pp. 50–56.

Sanchez-Pi, N., Orosa, L.M., Garcia, A.C.B., Information extraction techniques for health, safety, and environment applications in oil industry, *Proceedings of the IADIS International Conference on Intelligent Systems and Agents*, 2013, pp. 115–117.

Santos-Reyes, J., Beard, A.N., A SSMS model with application to the oil and gas industry, *Journal of Loss Prevention in the Process Industries*, 22, 2009, 958–970.

Santos-Reyes, J., Santos-Reyes, D., Assessment of safety management systems in the oil and gas industry, *Proceedings of the Offshore Technology Conference*, 2002, pp. 1469–1485.

Satin, K., Stock, A., Conducting effective health assessments in the oil and gas industry, *Proceedings of the SPE International Conference on Health, Safety, and Environment in Oil and Gas Exploration and Production*, 2010, pp. 2059–2067.

Schunder-Tatzber, S., Unterberger, W., Health circles—A tool for work place health promotion in the oil industry, *Proceedings of the 9th International Conference on Health, Safety, and Environment in Oil and Gas Exploration and Production*, 2008, pp. 1926–1939.

Seaton, A., Hazards to health in the shale oil industry: Lessons from Scotland, *Proceedings of the 4th Annual RMCOEH Occupational and Environmental Health Conference*, 1983, pp. 437–442.

Serwinowski, M.A., Marshall, J., The ROI of social responsibility: Driving sustainability in the oil gas sector, *Proceedings of the SPE International Conference on Health, Safety, and Environment in Oil and Gas Exploration and Production,* 2010, pp. 2725–2731.

Sexton, K., Visser, K., Health, safety, and environmental auditing in EP industry, *Proceedings of the 1st International Conference on Health, Safety, and Environment in Oil and Gas Exploration and Production,* 1991, pp. 731–734.

Shikdar, A.A., Sawaqed, N.M., Ergonomics, and occupational health and safety in the oil industry: A manager's response, *Computers and Industrial Engineering,* 47, 2004, 223–232.

Simmerman, J.A., Challenges of soft target security protection in the oil and gas industry, *Proceedings of the SPE Americas E and P Health, Safety, Security, and Environmental Conference,* 2013, pp. 166–168.

Simons, H.F., Safety attitude of oil industry resulting in fewer accidents, *Oil and Gas Journal,* 39, 1940, 45–49.

Sivandran, S., Risk and reliability analyses for driving design improvement in offshore engineering, *Proceedings of the 21st International Offshore and Polar Engineering conference,* 2011, pp. 756–761.

Skogdalen, J.E., Utne, I.B., Vinnem, J.E., Developing safety indicators for preventing offshore oil and gas deepwater drilling blowouts, *Safety Science,* 49, 2011, 1187–1199.

Skogdalen, J.E., Vinnem, J.E., Quantitative risk analysis offshore—Human and organizational factors, *Reliability Engineering and System Safety,* 96, 2011, 468–479.

Smiley, T.F., Prevention rather than cure is aim, *Oil and Gas Journal,* 26, 1928, 40–106.

Smith, D., Safety performance of the global EP industry 2002, *Proceedings of the Offshore Technology Conference,* 2004, pp. 218–223.

Song, G., Patil, D., Kocurek, C., Bartos, J., Applications of shape memory alloys in offshore oil and gas industry: A review, *Proceedings of the 12th International conference on Engineering, Science, Construction, and Operations in Challenging Environments,* 2010, pp. 1551–1567.

Sortland, G.W., Madaboosi, M., Fox, S.E., Capturing management information in the oil and gas industry, *Proceedings of the 9th International Conference on Health, Safety, and Environment in Oil and Gas Exploration and Production,* 2008, pp. 874–879.

Speck, J.B., Iravani, A.T.M., Industry survey of risk-based life management practices, *Proceedings of the ASME Pressure Vessels and Piping Conference,* 2002, pp. 109–116.

Speiser-Rankine, N. et al., Development and implementation of process-oriented skin safety standards for the mineral oil industry: A pilot study, *Proceedings of the 8th SPE International Conference on Health, Safety, and Environment in Oil and Gas Exploration and Production,* 2006, pp. 853–863.

Sphaier, S.H. et al., On the safety of off-loading operation in the oil industry, *International Journal of Computer Applications in Technology,* 43, 2012, 207–216.

Spotts, R.W., Photovoltaic applications in the oil and gas industry instrumentation field, *Proceedings of the ISA International Conference and Exhibit,* 1984, pp. 563–565.

Strobhar, D.A., Human factors issues in oil and chemical processing, *Proceedings of the 8th International Topical Meeting on Nuclear Plant Instrumentation, Control, and Human–Machine Interface Technologies,* 2012, pp. 557–562.

Sui, D. et al., Ensemble methods for process monitoring in oil and gas industry operations, *Journal of Natural Gas Science and Engineering*, 3, 2011, 748–753.

Sutherland, K.M., Flin, R.H., Psychosocial aspects of the offshore oil industry, *Proceedings of the 1st International Conference on Health, Safety, and Environment in Oil and Gas Exploration and Production*, 1991, pp. 793–802.

Sykes, R., Depraz, S., Buckley, H., Partnerships in the oil and gas industry: An IPIECA perspective, *Proceedings of the 9th International Conference on Health, Safety, and Environment in Oil and Gas Exploration and Production*, 2008, pp. 1891–1900.

Teresa, U.M., Cardiovascular risk impact in the oil industry, *Proceedings of the SPE Latin American and Caribbean Health, Safety, Environment, and Social Responsibility Conference*, 2013, pp. 185–191.

Tian, H., Huang, Y., He, J., Liu, P., Qu, W., Application of fault tree analysis in the reliability analysis of oil-gas long pipeline, *Proceedings of the International Conference on Pipelines and Trenchless Technology*, 2013, pp. 1436–1447.

Van de Poel, I.R., Hale, A.R., Goossens, L.H.J., Safety management in dutch oil and gas industry: The effect on the technological regime, *International Journal of Technology, Policy and Management*, 2, 2002, 407–433.

Vasquez Cordano, A.L., Salvador Jacome, J., Garcia Carpio, Fernandez Guzman, V., Assessing risks and regulating safety standards in the oil and gas industry: The Peruvian experience, *Proceedings of the SPE Latin American and Caribbean Health, Safety, Environment, and Social Responsibility Conference*, 2013, pp. 160–175.

Vestly Bergh, L.I., Hinna, S., Leka, S., Jain, A., Developing a performance indicator for psychosocial risk in the oil and gas industry, *Safety Science*, 62, 2014, 98–106.

Vinnem, J.E., Red, W., Norwegian oil and gas industry project to reduce the number of hydrocarbon leaks with emphasis on operational barriers improvement, *Proceedings of the SPE European HSE Conference and Exhibition*, 2013, pp. 307–317.

Vinogradov, V.N., Ibragimov, I.A., Special features of training engineers for the gas and oil industry at the present stage of development, *Proceedings of the 11th World Petroleum Congress*, 1984, pp. 31–37.

Wagenaar, W.A., Groeneweg, J., Hudson, P.T.W., Reason, J.T., Promoting safety in the oil industry, *Ergonomics*, 37, 1994, 1999–2013.

Wallace, I.G., Safety auditing in the offshore industry, *Proceedings of the Institution of Chemical Engineers Symposium*, 1990, pp. 85–97.

Wang, T., Xuan, W., Wang, X., Ren, K., Overview of oil and gas pipeline failure database, *Proceedings of the International Conference on Pipelines and Trenchless Technology*, 2013, pp. 1161–1167.

Wang, W., Majid, H.B.A., Reliability data analysis and modelling of offshore oil platform plant, *Journal of Quality in Maintenance Engineering*, 6, 2000, 287–295.

Webster, A., Condition monitoring process for critical rotating equipment within the oil and gas industry, *Proceedings of the 10th European Fluid Machinery Congress*, 2008, pp. 201–210.

Yang, H.D., Xu, H., Reliability analysis of gas turbine based on the failure mode and effect analysis, *Proceedings of the Asia-Pacific Power and Energy Conference*, 2011, pp. 1–4.

Yang, X., Sam Mannan, M., The development and application of dynamic operational risk assessment in oil/gas and chemical process industry, *Reliability Engineering and System Safety*, 95, 2010, 806–815.

Zappador, C., Gomez, G., Madera, A., Uberti, F., Health reporting system in the oil gas business, *Proceedings of the 13th Abu Dhabi International Petroleum Exhibition and Conference*, 2008, pp. 1509–1516.

Zhang, H., Gao, D., Evaluation of oil and gas drilling technology with safety analysis, *Proceedings of the International Symposium on Safety Science and Technology*, 2004, pp. 463–467.

Zhang, P., Zhou, Y., Liao, T., Research on quantitative risk assessment model and failure probability of oil/gas pipeline based on dempster-shafer evidence and intuitionist fuzzy theory, *Proceedings of the International Conference on Quality, Reliability, Risk, Maintenance, and Safety Engineering*, 2013, pp. 145–150.

Index

Milton Keynes UK
Ingram Content Group UK Ltd.
UKHW040102071024
449327UK00019B/734